CliffsNotes® Chemistry Quick Review

By Harold D. Nathan, Ph.D. and
Charles Henrickson, Ph.D.

Revised by Robyn L. Ford, M.Ed.

3rd Edition

About the Authors

Charles Henrickson received the PhD in chemistry from the University of Iowa with additional work at the University of Illinois and the University of California, Berkeley. He currently is Professor of Chemistry at Western Kentucky University.

Robyn L. Ford is a classroom teacher having taught all levels of chemistry and physics. She currently teaches a dual enrollment course in AP Chemistry and adjuncts at a local university. She is working toward her Ph.D. in Chemistry Education.

Publisher's Acknowledgments

Editorial

Acquisitions Editor: Greg Tubach
Project Editor: Kelly D. Henthorne
Technical Editors: Dr. Lester L. Pesterfield, Ph.D., and Dr. Darwin Dahl, Ph.D.

Composition

Indexer: BIM Indexing & Proofreading Services
Proofreader: Henry Lazarek
Wiley Publishing, Inc. Composition Services

CliffsNotes® Chemistry Quick Review, 3rd Edition
Published by:
Learneo, Inc.
www.learneo.com

Library of Congress Control Number: 2011927302

ISBN: 978-1-957671-64-2

Printed in the United States of America
10 9 8 7 6 5 4 3 2

Table of Contents

Introduction **1**
Why Do You Need This Book? 1
How to Use This Book 1
Hundreds of Practice Questions Online! 2

Chapter 1: Elements **3**
Discovery and Similarity 3
Atomic Masses 5
The Periodic Table 7

Chapter 2: Atoms **13**
Chemical Compounds 14
Stoichiometry 15
The Mole Unit 17
Chemical Reactions 18

Chapter 3: Atomic Structure **21**
Subatomic Particles 22
Isotopes 23
Radioactivity 25
Ions 27

Chapter 4: Electron Configurations **31**
Orbitals 31
Valence Electrons 33
The Periodic Table 35

Chapter 5: Chemical Bonding **39**
Covalent Bonds 40
Ionic Bonds 43
Polar Bonds 47
Other Bonds 49

Chapter 6: Organic Compounds **53**
Structural Formulas 54
Hydrocarbons 56
Compounds with Additional Elements 60

Chapter 7: States of Matter **65**
Solids, Liquids, and Gases 66
Phase Diagrams 66
Heat Capacities and Transformations 71

Chapter 8: Gases **75**
Boyle's Law 75
Charles' Law 77
Avogadro's Law 78
Ideal Gas Equation 79

Chapter 9: Solutions **83**
Concentration Units 84
Solubility 87
Freezing and Boiling Points 91

Chapter 10: Acids and Bases **97**
The pH Scale 98
Strong and Weak Acids 100
Two Types of Bases 103
Polyprotic Acids 104

Chapter 11: Oxidation-Reduction Reactions **107**
Oxidation Numbers 108
Electron Transfer 109
Balancing Equations 110

Chapter 12: Electrochemistry **115**
Electrochemical Cells 115
Electrode Potential 117
Faraday's Laws 120

Chapter 13: Equilibrium **125**
Two Reaction Directions 126
Equilibrium Concentrations 128
Le Chatelier's Principle 132

Chapter 14: Thermodynamics **137**
Enthalpy 138
Energy and Entropy 141
Gibbs Free Energy 144

Review Questions **151**

Resource Center **161**
Books 161
Internet 161

Glossary **163**

Appendix **169**

Index **179**

Introduction

You can learn chemistry! It's not impossible if you start right at the beginning—learning about the elements, the building blocks of nature. That's how you do it in this book. You start with the elements and learn that every element exists as atoms. Learning what these atoms look like and how they join together to form molecules and compounds is the next step, and along the way, you will gain an understanding of the way nature works. From that point, it is a matter of looking at different kinds of compounds and elements and learning how their properties make each of them unique. You'll start with the elements and end by seeing how batteries work, with a bunch of interesting stuff in between.

Although it is useful if you have some general science and math in your background, it isn't mandatory. And even if you choose to overlook most of the calculations in the book, you can still learn a great deal of chemistry. Don't try to go too fast. Take your time; you will learn chemistry!

Why Do You Need This Book?

Can you answer yes to any of these questions?

- Do you need to review the fundamentals of chemistry fast?
- Do you need a course supplement to chemistry?
- Do you need a concise, comprehensive reference for chemistry?

If so, CliffsNotes *Chemistry Quick Review* is for you!

How to Use This Book

You are the boss here. You get to decide how to use this book. You can either read the book from cover to cover or just look for the information you want and put the book back on the shelf for later. However, here are a few recommended ways to search for topics.

- Look for areas of interest in the book's table of contents or use the index to find specific topics.
- Flip through the book, looking for subject areas at the top of each page.

- Get a glimpse of what you'll gain from a chapter by reading through the "Chapter Check-In" at the beginning of each chapter.
- Use the "Chapter Check-Out" at the end of each chapter to gauge your grasp of the important information you need to know.
- Challenge yourself with the "Problems" sprinkled throughout the chapters. The answer explanations can be found in the appendix at the back of the book.
- Test your knowledge more completely in the "Review Questions" and look for additional sources of information in the "Resource Center."
- Use the glossary to find key terms fast. This book defines new terms and concepts where they first appear in the chapter. If a word is boldfaced, you can find a more complete definition in the book's glossary.
- Or flip through the book until you find what you're looking for—we organized this book to gradually build on key concepts.

Hundreds of Practice Questions Online!

Prepare for your next Chemistry quiz or test with hundreds of additional practice questions online. The questions are organized by this book's chapter sections, so it's easy to use the book and then quiz yourself online to make sure you know the subject. Go to www.cliffsnotes.com/go/quiz/chemistry to test yourself anytime and find other free homework help.

Chapter 1
ELEMENTS

Chapter Check-In

- ❑ Discovering the building blocks of all substances
- ❑ Understanding the arrangement of the periodic table
- ❑ Learning about atomic number and atomic weight

With all the different substances that exist, you may be surprised to learn that they are formed from a relatively small number of elements. For example, carbon is one of 112 known elements. Yet carbon can combine with other elements (like hydrogen and oxygen) to form thousands of substances (like sugar, alcohol, and plastics). Although some elements have been known from the earliest times, most were discovered during the last 300 years.

Each element is unique with its own characteristics. Each element is represented conveniently by a symbol. For example: H is for hydrogen, O for oxygen, and Cl for chlorine. In addition, each element has its own atomic number and atomic weight. The symbols of the elements, along with their respective atomic numbers and atomic weights, are displayed in a special arrangement called the **periodic table.** Although the elements are arranged in order of increasing atomic number, the elements with similar chemical characteristics appear in columns. This organization is useful for studying chemistry.

Discovery and Similarity

The modern science of chemistry began during the eighteenth century, when several brilliant natural philosophers classified the products of **decomposition** into a small number of fundamental substances. For example, in 1774, the Englishman Joseph Priestley discovered that when the red powder mercuric oxide was heated, it decomposed to liquid metal mercury and to a colorless gas capable of supporting combustion. (This gas was later named oxygen.) Most substances similarly can be decomposed into several simpler substances by heat or by an electrical current; however, the most fundamental substances cannot be broken down further, even with extraordinary temperature or electric voltage. These basic building blocks of all other substances are known as the chemical **elements.**

When the French chemist Antoine Lavoisier published his famous list of elements in 1789, there were only 33 elements, several of which were erroneous. By 1930, the diligent labors of thousands of chemists had increased the tally of naturally occurring chemical elements to 90. More recently, physicists in high-energy laboratories have been able to create about 20 highly radioactive, unstable elements that do not exist naturally on Earth, although they are probably produced in the hot cores of some stars.

The number of chemical elements has now reached 112, and the list is growing. Fortunately for students, only about 40 are relevant to basic chemistry. Please take a glance at the periodic table of chemical elements (found at the end of this chapter) and find calcium, element number 20. You need to be acquainted with the **symbol** and general properties of the 20 simplest elements up to calcium, plus another 20 of chemical significance that you will encounter in this book.

Notice that the key concepts of chemistry are set in boldface on their first appearance in the text to alert you to their importance. These terms are used repeatedly in this book, and you cannot master chemistry without understanding them. You can find the definitions of these key concepts in the glossary at the end of this book.

Groups of elements in nature have similar chemical properties. Helium, neon, argon, krypton, and xenon are all colorless gases, only two of which combine with other elements under very special conditions; their lack of reactivity leads to the name **inert gases** (or **noble gases**) for this group of similar elements. By contrast, fluorine and chlorine are corrosive greenish gases that form salts when they readily combine with metals, hence the name **halogens** (salt formers) for fluorine, chlorine, bromine, and iodine.

As a final example of a group of elements with similar properties, the metallic elements lithium, sodium, and potassium have such low densities that they float on water and are so highly reactive that they spontaneously burn by extracting oxygen from the water itself. These light metals form strong alkalis and appropriately are called the **alkali metals.** You should locate each of these columns of similar elements, as shown in Figure 1-1, on the periodic table.

Figure 1-1 Three groups of similar elements.

3 Li Lithium 6.94		2 He Helium 4.00
11 Na Sodium 22.99	9 F Fluorine 19.00	10 Ne Neon 20.18
19 K Potassium 39.10	17 Cl Chlorine 35.45	18 Ar Argon 39.95
37 Rb Rubidium 85.47	35 Br Bromine 79.90	36 Kr Krypton 83.80
55 Cs Cesium 132.91	53 I Iodine 126.90	54 Xe Xenon 131.30
Alkali metals	The halogens	Inert gases

Similar elements also occur in the same natural environment. For instance, the halogens are markedly concentrated in seawater. (The major salt in ocean brines is sodium chloride.) The other halogens are extracted from seawater that has been further concentrated—bromine from salt beds formed by evaporation and iodine from kelp, which grows in oceans.

The first indispensable key to making sense out of the extensive system of facts and principles called *chemistry* is the rule that the behavior of an element or compound can be predicted from similar substances.

Atomic Masses

By the early nineteenth century, chemists were striving to organize their rudimentary knowledge of the chemical elements. It was known that differing masses of elements reacted to form compounds. For example, they found that 3 grams of magnesium metal reacted with precisely 2 grams of oxygen to form magnesium oxide with no residual magnesium or oxygen. The same mass of oxygen, however, required 5 grams of calcium metal to react completely to form calcium oxide. Table 1-1 summarizes these relative combining masses.

Table 1-1 Ratios from Experiments

Combining Masses
5 grams calcium
3 grams magnesium
2 grams oxygen

Chemists gradually discovered that such relative masses in chemical reactions were fundamental characteristics of the elements. The English chemist John Dalton realized that all the known combining masses were nearly whole-number multiples of the combining mass of the lightest element—hydrogen. In 1803, he proposed an atomic theory in which all other elements would be built from multiple hydrogen atoms. Consequently, he based his scale of **atomic masses** on hydrogen being equal to 1.

Although Dalton's theory was found to be unrealistically simple, he did compel chemists to adopt a standard scale of atomic weights. Because the combining mass of oxygen is approximately 16 times that of hydrogen, the preceding chart can be revised, as shown in Table 1-2.

Table 1-2 Standard Scale

Atomic Masses
40.08 calcium
24.31 magnesium
16.00 oxygen
1.01 hydrogen

The modern masses for calcium, magnesium, and oxygen are still nearly in the 5:3:2 ratios of the original masses. Notice especially that the atomic mass of hydrogen is not precisely equal to 1, because the atomic mass scale is now based on the most common variety of carbon being exactly

12 atomic mass units. Dalton's bold conjecture that all the heavier elements have masses that are integral multiples of hydrogen is not strictly valid, but his theory was a good approximation that eventually led to the discovery of the particles composing the atoms.

The Periodic Table

In 1869, the Russian chemist Dmitri Mendeleyev published his great systematization called the periodic table. He arranged all known chemical elements in order of their atomic masses and found that similar physical and chemical properties recurred every 7 elements for the lighter elements and every 17 elements for the heavier ones. (The inert gases had not been discovered at that time; the correct values for similar properties are 8 and 18.) The periodic table is based on atomic masses and similar properties. In each row, the atomic masses increase toward the right. Each column contains a group of elements with similar chemical behavior.

In the modern periodic table, each box contains four data, as shown in Figure 1-2. Besides the element name and symbol, the atomic mass is at the bottom, and the atomic number is at the top. The elements are arranged in order of increasing atomic number in horizontal rows called **periods.**

Figure 1-2 Facts given for each element.

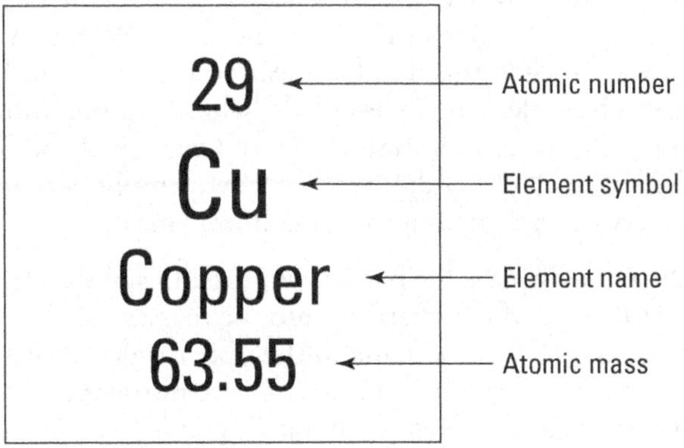

In the preceding section, you reviewed the historical determination of atomic masses. Notice that the elements also seem to be arranged in order of increasing atomic mass. But several exceptions exist. Compare the atomic mass of tellurium (Te) to iodine (I). (See Figure 1-3.)

Figure 1-3 Comparing tellurium and iodine.

16 S Sulfur 32.06	17 Cl Chlorine 35.45
34 Se Selenium 78.96	35 Br Bromine 79.90
52 Te Tellurium 127.60	53 I Iodine 126.90

Consider the proper placement of tellurium and iodine in the periodic table, as shown in Figure 1-3. Te has the heavier atomic mass. The chemical properties of tellurium are like those of selenium because both are semimetallic elements that form compounds like those of sulfur. Iodine resembles bromine because these elements are nonmetallic halogens that form compounds like those of chlorine. Therefore, the order in the table cannot be based solely on atomic mass.

The **atomic number,** which appears above each element symbol, represents the meaningful order in the periodic table. When an element is referred to by an integer, this number means the atomic number, not the atomic mass. Thus, element 27 is cobalt (whose atomic number is 27), not aluminum (whose atomic mass is 27). In Chapters 2 and 3, these two concepts are more carefully defined; for now, simply bear in mind the distinction between atomic number and atomic mass.

The periodic table displays the pattern of properties of the elements. The lightest are at the top of the chart; the atomic masses increase toward the bottom of the chart. The elements to the upper right, above a diagonal line from aluminum (13) to polonium (84), are **nonmetals,** about half of which exist as gases under normal laboratory conditions. All the elements in the middle and left of the table are **metals,** except gaseous hydrogen (1). Most of the metals are shiny, deformable solids, but mercury has such a low melting point that it is a liquid at room temperature. All the metals have high conductivities for heat and electricity. Many simple chemical compounds are formed from a metal reacting with a nonmetal.

In the periodic table, shown on the following pages, elements in columns have similar properties, and elements so related (like sulfur, selenium, and tellurium) are members of the same **group** or family and are **congeners** of one another.

1 H Hydrogen 1.01								
3 Li Lithium 6.94	4 Be Beryllium 9.01							
11 Na Sodium 22.99	12 Mg Magnesium 24.31							
19 K Potassium 39.10	20 Ca Calcium 40.08	21 Sc Scandium 44.96	22 Ti Titanium 47.90	23 V Vanadium 50.94	24 Cr Chromium 52.00	25 Mn Manganese 54.94	26 Fe Iron 55.85	27 Co Cobalt 58.93
37 Rb Rubidium 85.47	38 Sr Strontium 87.62	39 Y Yttrium 88.91	40 Zr Zirconium 91.22	41 Nb Niobium 92.91	42 Mo Molybdenum 95.94	43 Tc Technetium (99)	44 Ru Ruthenium 101.07	45 Rh Rhodium 102.91
55 Cs Cesium 132.91	56 Ba Barium 137.34	57 La Lanthanum 138.91	72 Hf Hafnium 178.49	73 Ta Tantalum 180.95	74 W Tungsten 183.85	75 Re Rhenium 186.21	76 Os Osmium 190.2	77 Ir Iridium 192.22
87 Fr Francium (223)	88 Ra Radium (226)	89 Ac Actinium (227)	104 Rf Rutherfordium (257)	105 Db Dubnium (260)	106 Sg Seaborgium (263)	107 Bh Bohrium (262)	108 Hs Hassium (265)	109 Mt Meitnerium (266)

Lanthanides	58 Ce Cerium 140.12	59 Pr Praseodymium 140.91	60 Nd Neodymium 144.24	61 Pm Promethium (147)	62 Sm Samarium 150.35	63 Eu Europium 151.96
Actinides	90 Th Thorium (232)	91 Pa Protactinium (231)	92 U Uranium (238)	93 Np Neptunium (237)	94 Pu Plutonium (242)	95 Am Americium (243)

									2 He Helium 4.00
			5 B Boron 10.81	6 C Carbon 12.01	7 N Nitrogen 14.01	8 O Oxygen 16.00	9 F Fluorine 19.00	10 Ne Neon 20.18
			13 Al Aluminum 26.98	14 Si Silicon 28.09	15 P Phosphorus 30.97	16 S Sulfur 32.06	17 Cl Chlorine 35.45	18 Ar Argon 39.95
28 Ni Nickel 58.71	29 Cu Copper 63.55	30 Zn Zinc 65.38	31 Ga Gallium 69.72	32 Ge Germanium 72.59	33 As Arsenic 74.92	34 Se Selenium 78.96	35 Br Bromine 79.90	36 Kr Krypton 83.80
46 Pd Palladium 106.42	47 Ag Silver 107.87	48 Cd Cadmium 112.41	49 In Indium 114.82	50 Sn Tin 118.69	51 Sb Antimony 121.75	52 Te Tellurium 127.60	53 I Iodine 126.90	54 Xe Xenon 130.30
78 Pt Platinum 195.09	79 Au Gold 196.97	80 Hg Mercury 200.59	81 Tl Thallium 204.37	82 Pb Lead 207.19	83 Bi Bismuth 208.98	84 Po Polonium (210)	85 At Astatine (210)	86 Rn Radon (222)
110 Ds Darmstadtium (281)	111 Rg Roentgenium (280)	112 Cn Copernicium (285)						

64 Gd Gadolinium 157.25	65 Tb Terbium 158.93	66 Dy Dysprosium 162.50	67 Ho Holmium 164.93	68 Er Erbium 167.26	69 Tm Thulium 168.93	70 Yb Ytterbium 173.04	71 Lu Lutetium 174.97
96 Cm Curium (247)	97 Bk Berkelium (247)	98 Cf Californium (251)	99 Es Einsteinium (254)	100 Fm Fermium (257)	101 Md Mendelevium (258)	102 No Nobelium (259)	103 Lr Lawrencium (260)

This information was obtained from the International Union of Pure and Applied Chemistry (IUPAC) Web site: www.IUPAC.org.

Chapter Check-Out

1. The number of known elements is approximately ______.
   a. 44
   b. 112
   c. 520

2. Elements aligned in columns on the periodic table ______.
   a. have similar atomic weights
   b. have similar atomic numbers
   c. have similar chemical properties

3. The atomic number of iron (Fe) is ______.
   a. 26
   b. 53
   c. 55
   d. 85

4. Element number 4 is ______.
   a. helium (He)
   b. beryllium (Be)
   c. carbon (C)

5. Ultimately, the periodic table arranges elements from left to right in order of ______.
   a. the first letter of the element's name
   b. increasing atomic numbers
   c. increasing atomic weights

Answers: **1.** b **2.** c **3.** a **4.** b **5.** b

Chapter 2
ATOMS

Chapter Check-In

- ❑ Learning about atoms
- ❑ Forming compounds with combined atoms
- ❑ Determining formulas of compounds
- ❑ Learning the importance of the mole in chemistry
- ❑ Writing chemical reactions

The simplest unit of an element is the **atom.** The atom is never broken down into smaller particles in chemical reactions. The relative mass of an atom is given by its atomic mass, which appears on the periodic table for each element. An amount of an element equal to its atomic mass in grams is one **mole** of the element.

Atoms combine in definite whole number ratios to form compounds. Water (H_2O) is composed of two atoms of hydrogen (H) and one atom of oxygen (O). One mole of water is about 18.0 grams of water, an amount equal to the sum of the atomic masses of two hydrogen atoms and one oxygen atom in grams: 1.00 g + 1.00 g + 16.0 g = 18.0 g. Speaking in terms of moles is common when describing amounts of substances.

The chemical reaction between two or more substances is described using a **chemical equation.** The reaction between hydrogen and oxygen to form water is shown as

$$2H_2\ (g) + O_2\ (g) \rightarrow 2H_2O\ (l)$$

Chemical Compounds

The whole-number ratios of combining masses in chemical reactions can be readily explained knowing that the basic unit of all elements is the atom, which originally meant an indivisible particle. Therefore, one atom of carbon can react with either one atom of oxygen,

$$\underset{\text{carbon atom}}{\mathrm{C}} + \underset{\text{oxygen atom}}{\mathrm{O}} \rightarrow \underset{\text{carbon monoxide molecule}}{\mathrm{CO}}$$

or two atoms of oxygen,

$$\underset{\text{carbon atom}}{\mathrm{C}} + \underset{\text{oxygen molecule}}{\mathrm{O_2}} \rightarrow \underset{\text{carbon dioxide molecule}}{\mathrm{CO_2}}$$

but not with, say, $1\frac{1}{2}$ oxygen atoms. The reactions of carbon with oxygen are shown using chemical equations with reactants on the left and products on the right of the arrow.

A substance containing atoms of more than one element in a definite ratio is called a **compound.** The composition of a compound is shown in its **chemical formula.** In the chemical reaction of carbon and oxygen to form carbon dioxide, the elements are in a definite 1:2 ratio—one atom of carbon with two atoms of oxygen forming the compound carbon dioxide. When several atoms are so tightly bonded together that they physically behave as a unit, the unit is called a **molecule.** (See Figure 2-1.) Elements and compounds can be molecular. In the carbon dioxide reaction, the O_2 molecule contains 2 atoms of oxygen, and the CO_2 molecule has 3 atoms—1 carbon and 2 oxygen.

Figure 2-1 Atoms and molecules.

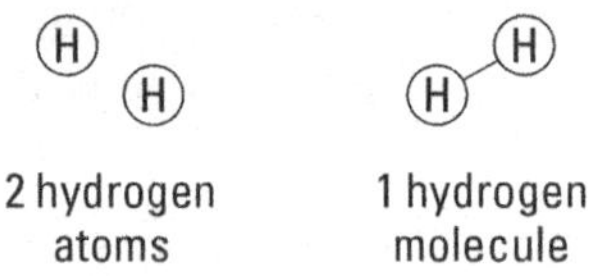

Five major compounds are built solely of nitrogen and oxygen, as shown in Table 2-1. Of the five oxides of nitrogen, nitrous oxide is richest in nitrogen, and dinitrogen pentoxide is richest in oxygen.

Table 2-1 The Oxides of Nitrogen

Compound Name	*Chemical Formula*	*Atomic Ratio (N:O)*
Nitrous oxide	N_2O	2:1
Nitric oxide	NO	1:1
Dinitrogen trioxide	N_2O_3	2:3
Nitrogen dioxide	NO_2	1:2
Dinitrogen pentoxide	N_2O_5	2:5

Stoichiometry

The atomic ratios in each compound are also the relative number of atomic mass units of its elements. The first example is nitrous oxide (N_2O), as shown in Table 2-2.

Table 2-2 Nitrous Oxide

Chemical Element	*Atomic Ratio*	*Atomic Mass*	*Relative Mass*	*Mass Percent*
Nitrogen	2	14.01	28.02	63.65
Oxygen	1	16.00	16.00	36.35

The relative masses were obtained by multiplying the atomic ratios and atomic masses. You can see that a sample of N_2O weighing 44.02 grams contains 28.02 g of nitrogen and 16.00 g of oxygen. The mass percent of each element is calculated from its relative mass divided by the sum of the relative masses. Chemical compounds with integral atomic ratios, like nitrous oxide, are described as **stoichiometric** compounds, and they permit many simple calculations.

The common oxide of aluminum provides a second example, but this time, begin with the mass percent and deduce the atomic ratio. Careful laboratory analysis of aluminum oxide determines it to be approximately 53 percent aluminum and 47 percent oxygen by mass, as shown in the second column in Table 2-3.

Table 2-3 Aluminum Oxide

Chemical Element	*Atomic Ratio*	*Atomic Mass*	*Relative Mass*	*Mass Percent*
Aluminum	2	26.98	1.916	52.92
Oxygen	3	16.00	2.943	47.08

The relative masses are obtained by dividing the mass percentages by the atomic masses from the periodic table.

$$\text{aluminum } \frac{52.92}{26.98} = 1.961 \qquad \text{oxygen } \frac{47.08}{16.00} = 2.943$$

So, the ratio of oxygen to aluminum in relative atomic masses is

$$\frac{\text{oxygen}}{\text{aluminum}} = \frac{2.943}{1.961} = 1.50$$

If the aluminum oxide is a stoichiometric compound with whole-number ratios of the constituent elements, the preceding quotient must be translated into a ratio of integers:

$$1.50 = \frac{3 \text{ oxygen}}{2 \text{ aluminum}}$$

This ratio of integers implies that the formula for aluminum oxide is $A1_2O_3$. In simple compounds, the metallic elements are written before the nonmetals.

Most chemical compounds are stoichiometric, and you should be able to utilize the atomic masses and perform the following calculations:

- Mass percents from the chemical formula
- Chemical formula from the weight percents

These calculations are so basic to the field that you should go back and carefully review the two examples in this section: the calculation of mass percents in N_2O and the inference of the formula for aluminum oxide. Then you can practice stoichiometric calculations on the following pair of problems, which are answered and explained in the appendix. Many such practice exercises are included in this book so that you can determine whether you understand the major concepts of chemistry. It is well worth your time to study these examples and their explanations in the appendix until you can do the calculations correctly.

Problem 1: Determine the mass percentages of the three elements in ammonium chloride, which has the formula NH_4Cl.

Problem 2: Infer the simplest chemical formula for potassium copper (II) fluoride, which has this analysis by mass:

Potassium	35.91%
Copper	29.19%
Fluorine	34.90%

The Mole Unit

In laboratory practice, you work with much larger quantities of the elements than single atoms or single molecules. A convenient standard quantity is the **mole,** the amount of the substance equal in grams to the sum of the atomic masses. Therefore, one mole of carbon dioxide is 44.01 grams:

$$\underset{\text{carbon}}{C} + \underset{\text{oxygen}}{2(O)} = \underset{\text{carbon dioxide}}{CO_2}$$

$$12.01 + 2(16.00) = 44.01$$

The mole is a convenient unit for expressing the relative amounts of substances in chemical reactions. The burning of carbon in oxygen can be written with the 2 oxygen atoms bonded in a single O_2 molecule:

$$\underset{\substack{\text{1 mole carbon} \\ (12.01\text{ g})}}{C} + \underset{\substack{\text{1 mole molecular oxygen} \\ (32.00\text{ g})}}{O_2} \rightarrow \underset{\substack{\text{1 mole carbon dioxide} \\ (44.01\text{ g})}}{CO_2}$$

The mole is the most common unit used to express the quantity of a chemical substance. For all solids, liquids, and gases, you can convert mass to moles (or moles to mass). For gases, you should memorize the following conversion of volume to moles (or moles to volume): at 0°C and 1 atmosphere pressure (known as **standard temperature and pressure** or **STP**), one mole of any gas occupies approximately 22.4 liters. Therefore, the preceding reaction describing the oxidation of carbon means that 12 grams of carbon burned at STP in 22.4 liters of O_2 yields 22.4 liters of carbon dioxide.

Many substances do not exist as molecules. For example, the atoms in most inorganic solids are in a three-dimensional structure in which each atom is surrounded by a number of other atoms. In crystals of sodium chloride, no distinct Na and Cl pair can be called a molecule, because each sodium is surrounded by 6 chlorine and each chlorine is surrounded by 6 sodium.

For nonmolecular substances like sodium chloride, the use of the word mole, with its connotation of molecules, is inappropriate. A comparable unit, the **gram formula mass,** is used; it is defined as the sum in grams of the atomic masses of all the atoms in the chemical formula of the substance. For sodium chloride (NaCl), the gram formula mass is calculated as

$$\underset{\text{sodium}}{22.99\text{ g}} + \underset{\text{chlorine}}{35.45\text{ g}} = 58.44\text{ g NaCl}$$

The mole unit and the gram formula unit are employed in similar calculations. For this reason, many chemists use the term *mole* to describe quantities of both molecular and nonmolecular substances. Try answering the following problems:

Problem 3: How many moles of bromobenzene are in 1 kilogram of bromobenzene, C_6H_5Br?

Problem 4: What is the mass in grams of neon gas that occupies a volume of 5 liters at 0°C and 1 atmosphere of pressure?

Chemical Reactions

The standard representation of a chemical reaction shows an arrow pointing from the reactants to the products:

$$\underset{\text{iron (III) oxide}}{Fe_2O_3(s)} + \underset{\text{carbon monoxide}}{3CO(g)} \rightarrow \underset{\text{metallic iron}}{2Fe(s)} + \underset{\text{carbon dioxide}}{3CO_2(g)}$$

For most chemical reactions in this book, solids are labeled *(s)*, liquids *(l)*, and gases *(g)*. The numerical coefficients in front of the chemical formulas express the moles of each compound or element. The preceding reaction can be interpreted in terms of moles or masses. (See Table 2-4.)

Table 2-4 Interpretation of a Reaction

Quantity	Fe_2O_3	*CO*	*Fe*	CO_2
Moles	1	3	2	3
Mass of 1 mole	159.70	28.01	55.85	44.01
Total masses	159.70	84.03	111.70	132.03

Notice that the total mass of the reactants (243.73 g) equals the total mass of the **products.** This demonstrates the *law of conservation of mass,* which applies to all chemical reactions.

It is not true, however, that *volumes* must be conserved in reactions involving gases. The complete combustion of carbon monoxide is a case in point:

$$\underset{\text{carbon monoxide}}{2CO(g)} + \underset{\text{oxygen}}{O_2(g)} \rightarrow \underset{\text{carbon dioxide}}{2CO_2(g)}$$

Because the reaction coefficients are proportional to relative volumes of each gas, 2 volumes of carbon monoxide and 1 volume of oxygen (a total of 3 volumes of reactants) combine to produce only 2 volumes of carbon dioxide. In gaseous reactions, the total volume of the products may be less than, or equal to, or greater than the total volume of the reactants.

Problem 5: How many liters of oxygen (O_2) are required for the complete oxidation of 1 gram of methane when the following reaction occurs at STP?

$$\underset{\text{methane}}{CH_4(g)} + \underset{\text{oxygen}}{2O_2(g)} \rightarrow \underset{\text{carbon dioxide}}{CO_2(g)} + \underset{\text{water vapor}}{2H_2O(g)}$$

Chapter Check-Out

1. The smallest whole unit of an element is ______.

   a. a molecule
   b. an atom
   c. a mole

2. The chemical formula for methane, CH_4, shows that one molecule of methane contains ______.

   a. one atom of carbon (C) and four atoms of hydrogen (H)
   b. one gram of carbon and 4 grams of hydrogen
   c. one atom of carbon and one atom of hydrogen

3. In the compound nitric oxide (NO), for every 14 grams of nitrogen (N), there are how many grams of oxygen (O)?

   a. 16
   b. 8
   c. 14

4. In the following equation, how many moles of aluminum react with 3 moles of Cl_2?

   $2\ Al(s) + 3\ Cl_2(g) \rightarrow 2\ AlCl_3(s)$

   a. 1
   b. 2
   c. 3

5. The gram formula mass of calcium chloride, $CaCl_2$, weighs ______.

   a. 1 g
   b. 54 g
   c. 111 g

Answers: **1.** b **2.** a **3.** a **4.** b **5.** c

Chapter 3
ATOMIC STRUCTURE

Chapter Check-In

- ❑ Counting the particles that make up atoms
- ❑ Knowing the importance of atomic number
- ❑ Discerning differences in the isotopes of an element
- ❑ Learning the source of radioactivity
- ❑ Losing or gaining electrons forms an ion

You have already learned that atoms are the smallest units of elements. A block of iron is made up of a huge number of iron atoms packed together. In this chapter, you learn that atoms are also composed of still smaller particles of matter: protons, neutrons, and electrons. The neutrons and protons are packed together in a very small body called the nucleus, and the **electrons** exist in a diffuse cloud that completely encloses the much smaller nucleus. Atoms of the same element have the same number of protons in the nucleus but may differ by one or more neutrons forming **isotopes** of the element. Most elements exist in nature as mixtures of two or more isotopes. Because each isotope has its own atomic mass, the atomic mass reported for the element itself is the weighted average of its naturally occurring isotopes. Atoms can also gain or lose one or more electrons, forming charged species called **ions.**

If the number of neutrons and protons in a nucleus gets too large, the nucleus can become unstable (radioactive) and break apart, forming a new element, emitting small particles called **alpha** and **beta radiation,** and releasing energy called **gamma radiation.**

Subatomic Particles

The rather steady increase of atomic masses through the periodic table was explained when physicists managed to split atoms into three component particles.

The exploration of atomic structure began in 1911, when Ernest Rutherford, a New Zealander who worked in Canada and England, discovered that atoms had a dense central **nucleus** that contained positively charged particles, which he named **protons.** (See Table 3-1.) It was soon established that each chemical element was characterized by a specific number of protons in each atom. A hydrogen atom has 1 proton, helium has 2, lithium has 3, and so forth through the periodic table. The atomic number is the number of protons for each element.

Table 3-1 Parts of an Atom

Subatomic Particle	*Mass Units (amu)*	*Relative Electric Charge*	*Atomic Location*
Proton	1.0073	+1	Nucleus
Neutron	1.0087	0	Nucleus
Electron	0.0005	−1	Orbital

Except for the simplest hydrogen atom with a single proton as its entire nucleus, all atoms contain **neutrons** (particles that are electrically neutral) in addition to protons. For most of the light elements, the numbers of protons and neutrons in the nucleus are nearly equal. Table 3-2 shows the most common nucleus for each element with the atomic mass rounded to the nearest integer. You can see that the rounded-off atomic masses are the sum of the protons and neutrons for each atom. The sum of the protons and neutrons is the **mass number** of an atom.

Table 3-2 Nuclear Structure of the Lightest Elements

Element	*Atomic Number*	*Protons*	*Neutrons*	*Atomic Mass*
Hydrogen	1	1	0	1
Helium	2	2	2	4
Lithium	3	3	4	7
Beryllium	4	4	5	9

Element	*Atomic Number*	*Protons*	*Neutrons*	*Atomic Mass*
Boron	5	5	6	11
Carbon	6	6	6	12
Nitrogen	7	7	7	14
Oxygen	8	8	8	16

John Dalton's idea that atomic masses were multiples of hydrogen mass was premature, but near the truth. The series of elements of increasing atomic masses is generated by adding **nucleons,** the two types of particles comprising the nucleus, which are the protons and neutrons.

Isotopes

Although most of the lighter elements have atomic masses that are nearly whole numbers, some elements were discovered to have atomic masses that could not be integral. In Figure 3-1, look at the atomic masses of the three lightest halogens and satisfy yourself that although the values for fluorine and bromine may be whole numbers, the value of chlorine is definitely intermediate. The atomic mass of chlorine is not close to a whole number.

Figure 3-1 The three lightest halogens.

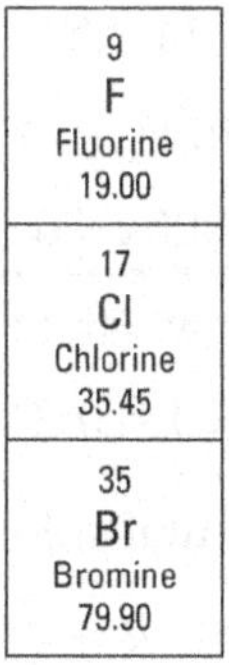

The interpretation of the curious mass of chlorine awaited the discovery of the neutron by James Chadwick in 1932. Although all chlorine atoms have 17 protons, different isotopes of the element have different numbers of neutrons. In Table 3-3, the mass numbers of the chlorine isotopes are denoted by superscripts to the upper left of the chemical symbol.

Table 3-3 Chlorine Isotopes

Atomic Isotope	*Number of Protons*	*Number of Neutrons*	*Atomic Mass*	*Relative Abundance*
^{35}Cl	17	18	34.97	76%
^{37}Cl	17	20	36.97	24%

The nonintegral atomic mass for naturally occurring chlorine is seen to be the weighted average of the atomic mass of its two major isotopes found by multiplying the atomic mass of each isotope times its decimal equivalent of its relative abundance:

$$0.76(\underset{^{35}\text{Cl}}{34.97}) + 0.24(\underset{^{37}\text{Cl}}{36.97}) = \underset{\text{natural mixture}}{35.45}$$

Now perform that calculation in the opposite direction. Beginning with the known atomic mass of natural chlorine, determine the abundance of the two isotopes:

$$x = \text{fraction } ^{37}\text{Cl}$$

$$1 - x = \text{fraction } ^{35}\text{Cl}$$

Instead of using the integers 37 and 35 as atomic masses, take the more precise atomic masses of the isotopes from Table 3-3:

$$36.97(x) + 34.97(1 - x) = 35.45$$

$$36.97x + 34.97 - 34.97x = 35.45$$

$$2.00x = 0.48$$

$$x = 0.24$$

$$1 - x = 1 - 0.24 = 0.76$$

The calculation reveals that natural chlorine is 24% ^{37}Cl and 76% ^{35}Cl.

The most carefully studied element is the simplest, hydrogen, which has a natural atomic mass (1.0080) slightly greater than that of a single proton (1.0078). (See Table 3-4.) This mass excess is only 0.0002 atomic mass units, but the investigation of this excess revealed the three isotopes of that element.

Table 3-4 Isotopes of Hydrogen

Natural Isotope	*Number of Protons*	*Number of Neutrons*	*Atomic Mass*	*Relative Abundance*
^{1}H	1	0	1.0078	99.985%
^{2}H	1	1	2.0141	0.015%
^{3}H	1	2	3.0161	Rare–negligible

^{2}H is often called deuterium, and ^{3}H is referred to as tritium. The atomic mass of natural hydrogen (1.0080) exceeds that of ^{1}H because of the admixture of deuterium:

$$\underset{^{1}\text{H}}{0.99985(1.0078)} + \underset{^{2}\text{H}}{0.00015(2.0141)} = \underset{\text{natural mixture}}{1.0080}$$

Problem 6: In the following chart, which nuclei are isotopes of one chemical element? Can you give the element's name? Which nuclei have nearly the same mass?

Nucleus	*Protons*	*Neutrons*
A	11	13
B	12	12
C	12	13

Problem 7: Listed in the following chart are the atomic masses (measured in atomic mass units) for natural silver and its two isotopes. Use this data to calculate the percentage of silver-109 in the natural mixture.

^{107}Ag	106.905
^{109}Ag	108.905
natural Ag	107.868

Radioactivity

If you look at the periodic table, you will notice that all elements after bismuth, atomic number 83, have their atomic mass denoted by an integer within parentheses. Such large nuclei are unstable and undergo **radioactivity,** the spontaneous disintegration by the emission of particles. This process is also known as **radioactive decay.** The atomic mass shown on the periodic table is the mass number of the most common isotope of each radioactive element.

For chemical purposes, the most important types of radiation are alpha and beta particles. An **alpha particle** (α) is a ${}^{4}_{2}He$ nucleus with 2 protons and 2 neutrons. This nuclear notation uses a subscript to the lower left to record the number of protons, whereas the superscript to the upper left is the *mass number,* the total number of nucleons. The number of protons identifies the chemical element, whereas the nucleon total is the mass number for the particle or isotope.

If an unstable nucleus emits an alpha particle, its atomic number decreases by 2 and its mass number decreases by 4. The decay of thorium-232 provides an example:

$$\underset{\text{thorium-232}}{{}^{232}_{90}Th} \rightarrow \underset{\text{radium-228}}{{}^{228}_{88}Ra} + \underset{\text{alpha particle}}{{}^{4}_{2}He}$$

Notice that radioactive decay actually changes one chemical element to another element, a process referred to as **transmutation**. Also notice that the total of the mass numbers on each side are the same, indicating they are conserved. This is also true for the total charge on each side.

The other chemically important mode of radioactive transmutation is provided by negative **beta particles** (β), which are electrons emitted by atomic nuclei, not from the surrounding electronic orbitals. The beta particle can be denoted as an electron (e^-) or with the Greek letter beta (${}^{0}_{-1}\beta$). The beta particle arises from the decay of a neutron to a proton:

$$\underset{\text{neutron}}{n^0} \rightarrow \underset{\text{proton}}{p^+} + \underset{\text{electron}}{{}^{0}_{-1}\beta}$$

The creation of the proton causes the atomic number to increase by one. An example of beta transmutation is the decay of lead-212:

$$\underset{\text{lead-212}}{{}^{212}_{82}Pb} \rightarrow \underset{\text{bismuth-212}}{{}^{212}_{83}Bi} + \underset{\text{beta particle}}{{}^{0}_{-1}\beta}$$

The atomic number increased to 83 due to the new proton, but the mass number stayed constant because both metal nuclei have 212 total protons plus neutrons.

The most familiar radioactive element is uranium, which has two naturally occurring isotopes of mass numbers 235 and 238 that decay very slowly. Review the first few steps in the decay of uranium-238, which changes to lead-206 after the emission of eight alpha and six beta particles. The earliest stages of the decay scheme involve only three elements, as shown in Figure 3-2.

Figure 3-2 Three radioactive elements.

90 Th Thorium (232)	91 Pa Protactinium (231)	92 U Uranium (238)

You should carefully examine the changes in both atomic number (protons) and mass number (protons plus neutrons):

$$^{238}_{92}\mathrm{U} \xrightarrow{\alpha} {}^{234}_{90}\mathrm{Th} \xrightarrow{\beta} {}^{234}_{91}\mathrm{Pa} \xrightarrow{\beta} {}^{234}_{92}\mathrm{U} \xrightarrow{\alpha} {}^{230}_{90}\mathrm{Th}$$

In this brief sequence, there are two different isotopes of uranium and two of thorium.

Problem 8: The next step in the uranium-238 decay scheme is the emission of an alpha particle from thorium-230. Describe the mass number, atomic number, and element name for the resulting nucleus.

Ions

Of the three major subatomic particles, the negatively charged electron was first discovered by the English physicist Joseph Thomson in 1897.

As the structure of atoms was probed, it was realized that these low-mass particles occurred in a large "cloud" around the tiny nucleus, which contained almost all the mass of the atom. A neutral atom has precisely equal numbers of protons (+) and electrons (–). Atoms with a charge imbalance are called ions. A positive ion has lost one or more electrons, whereas a negative ion has gained one or more electrons. Table 3-5 shows three different charge states for copper:

Table 3-5 Copper and Its Ions

Description	***Charge***	***Protons***	***Electrons***
Neutral atom	Cu^0	29	29
Cuprous ion	Cu^{1+}	29	28
Cupric ion	Cu^{2+}	29	27

The chemical behavior of the various elements is influenced more by the charge of their ion than by any other intrinsic property. In the periodic table (see Figure 3-3), elements in a column form analogous compounds because they have the same charge on their ions.

The halogens all tend to gain one electron, giving their ions a characteristic charge of –1. The **alkali metals,** on the other side of the periodic table, all readily lose one electron, so their ions possess a charge of +1. Charges are balanced in a compound with the number of alkali metal ions equal to the number of halogen ions; NaCl is such a compound. By contrast, consider the **alkaline earths** in the second column on the left side of the table. These elements are metals that form oxides that have an earthy texture and yield alkaline solutions. The alkaline earths have ions with a charge of +2. For one of them to form a compound with a halogen, twice as many halogen ions are needed to balance the +2 charge on the metal ions. Consequently, the correct chemical formula for strontium fluoride is SrF_2.

Figure 3-3 The alkali metals form ions with a +1 charge; the alkaline earths, +2; and the halogens, –1.

Alkali metals	Alkaline earths	The halogens
3 Li Lithium 6.94	4 Be Beryllium 9.01	9 F Fluorine 19.00
11 Na Sodium 22.99	12 Mg Magnesium 24.31	17 Cl Chlorine 35.45
19 K Potassium 39.10	20 Ca Calcium 40.08	35 Br Bromine 79.90
37 Rb Rubidium 85.47	38 Sr Strontium 87.62	53 I Iodine 126.90
55 Cs Cesium 132.91	56 Ba Barium 137.34	85 At Astatine (210)
+1	+2	–1

Ion charge

Chapter Check-Out

1. The subatomic particle that bears a negative charge is the ______.
   - a. neutron
   - b. proton
   - c. electron

2. The number of protons in the nucleus of an atom is its ______.
   - a. atomic number
   - b. atomic mass
   - c. isotope

3. Which atom is an isotope of iron?

	protons	neutrons	electrons
a.	25	31	25
b.	26	30	26
c.	55	29	26

4. What is the charge on an ion of sulfur if one sulfur atom gains two electrons?
   - a. –2
   - b. +2
   - c. 0

5. In radioactive decay, if an atom of uranium-238 emits an alpha particle, what element will be the product of the decay?
   - a. uranium (U)
   - b. thorium (Th)
   - c. plutonium (Pu)

Answers: **1.** c **2.** a **3.** b **4.** a **5.** b

Chapter 4
ELECTRON CONFIGURATIONS

Chapter Check-In

- ❑ Arranging electrons about the nucleus
- ❑ Writing electronic configurations of atoms and ions
- ❑ Configuring electrons to determine the properties of an element
- ❑ Learning which electrons are most important
- ❑ Determining the place of an element in the periodic table

In Chapter 3, you learned that protons and neutrons are packed in a very small nucleus buried in a cloud of electrons. Though the nucleus is not unimportant, the electrons determine how an element behaves chemically.

The electrons are arranged about the nucleus in specific regions called **shells.** The electrons of lower **energy** are held nearer the nucleus, and those of higher energy are placed farther away on the outer surface of the atom. The electrons in the highest energy shell, called the **valence electrons,** are the ones that most directly determine the chemistry of the element. Each shell is further broken down into subshells (*s*, *p*, *d*, and *f*), and each subshell is made up of 1, 3, 5, or 7 orbitals, which contain the electrons. Each orbital can contain 1 or 2 electrons, or it can be empty. Elements in the same vertical group in the periodic table all have their highest energy electrons arranged in the same way, and they are, therefore, chemically similar.

Orbitals

Quantum theory assigns the electrons surrounding the nucleus to **orbitals,** which should not be confused with the orbits of the solar system. Each orbital has a characteristic energy and a three-dimensional shape. An atom in the lowest energy configuration is said to be in its **ground state**.

For this most stable state, the electrons fill the various orbitals from the one of lowest energy to the one of highest energy. Each orbital may be assigned a maximum of two electrons.

The orbitals are described completely by specifying three quantum numbers, but only two are used in this book. The principal quantum number (symbolized n) is a whole number, 1 or greater, that identifies the main energy shell of the orbital, with 1 being closest to the nucleus and each subsequent level farther from the nucleus. The second quantum number, known as the azimuthal quantum number (symbolized ℓ) is a whole number from 0 up to $n - 1$ that defines the type of orbital within a shell (n). For historical reasons, the different shapes of orbitals are represented by letters. (See Table 4-1.)

Table 4-1 Types of Orbitals

Second Quantum Number	*Letter Denoting Orbitals*	*Number of Orbitals*	*Maximum Number Electrons*
0	*s*	1	2
1	*p*	3	6
2	*d*	5	10
3	*f*	7	14

Because each orbital holds at most two electrons, the maximum number of electrons is twice the number of orbitals for a particular second quantum number. In Table 4-1, you must know the letters in the second column and the electron capacity in the last column.

A set of orbitals with the same values n and ℓ is called a **subshell** and is represented by notation like $2p^5$. (See Figure 4-1.)

Figure 4-1 A subshell notation.

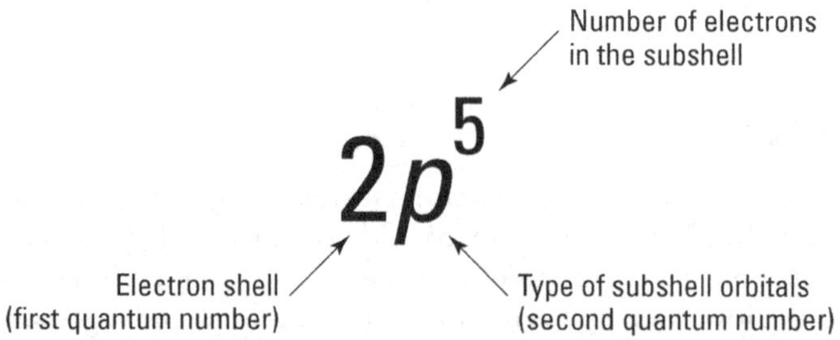

Only a few subshells are needed to describe the chemical elements in their ground states. Table 4-2 lists all the subshells of chemical importance.

Table 4-2 Electron Subshell

First Quantum Number	*Second Quantum Number*	*Notation for Subshells*
1	0	1*s*
2	0, 1	2*s*, 2*p*
3	0, 1, 2	3*s*, 3*p*, 3*d*
4	0, 1, 2, 3	4*s*, 4*p*, 4*d*, 4*f*
5	0, 1, 2, 3	5*s*, 5*p*, 5*d*, 5*f*
6	0, 1, 2	6*s*, 6*p*, 6*d*
7	0	7*s*

Valence Electrons

The electronic configuration of an atom is given by listing its subshells with the number of electrons in each subshell, as shown in Table 4-3. Study the third column of complete electronic configurations carefully so you understand how electrons are added to the subshell of lowest energy until it reaches its capacity; then the subshell of the next energy level begins to be filled. The electrons in the highest numbered subshells are the **valence** electrons, which comprise the valence shell of the atom.

Table 4-3 Electron Configurations and Oxidation Numbers

Element Name	*Atomic Number*	*Electron Configuration*	*Valence Shell*	*Common Oxidation Numbers*
Hydrogen	1	$1s^1$	$1s^1$	+1, −1
Helium	2	$1s^2$	$1s^2$	0
Lithium	3	$1s^2\ 2s^1$	$2s^1$	+1
Beryllium	4	$1s^2\ 2s^2$	$2s^2$	+2
Boron	5	$1s^2\ 2s^2\ 2p^1$	$2s^2\ 2p^1$	+3
Carbon	6	$1s^2\ 2s^2\ 2p^2$	$2s^2\ 2p^2$	+4, +2, −4

(continued)

Table 4-3 ***(continued)***

Element Name	*Atomic Number*	*Electron Configuration*	*Valence Shell*	*Common Oxidation Numbers*
Nitrogen	7	$1s^2\ 2s^2\ 2p^3$	$2s^2\ 2p^3$	+5, +3, –3
Oxygen	8	$1s^2\ 2s^2\ 2p^4$	$2s^2\ 2p^4$	–2
Fluorine	9	$1s^2\ 2s^2\ 2p^5$	$2s^2\ 2p^5$	–1
Neon	10	$1s^2\ 2s^2\ 2p^6$	$2s^2\ 2p^6$	0
Sodium	11	$1s^2\ 2s^2\ 2p^6\ 3s^1$	$3s^1$	+1
Magnesium	12	$1s^2\ 2s^2\ 2p^6\ 3s^2$	$3s^2$	+2
Aluminum	13	$1s^2\ 2s^2\ 2p^6\ 3s^2\ 3p^1$	$3s^2\ 3p^1$	+3
Silicon	14	$1s^2\ 2s^2\ 2p^6\ 3s^2\ 3p^2$	$3s^2\ 3p^2$	+4
Phosphorus	15	$1s^2\ 2s^2\ 2p^6\ 3s^2\ 3p^3$	$3s^2\ 3p^3$	+5, +3, –3
Sulfur	16	$1s^2\ 2s^2\ 2p^6\ 3s^2\ 3p^4$	$3s^2\ 3p^4$	+6, +4, +2, –2
Chlorine	17	$1s^2\ 2s^2\ 2p^6\ 3s^2\ 3p^5$	$3s^2\ 3p^5$	–1
Argon	18	$1s^2\ 2s^2\ 2p^6\ 3s^2\ 3p^6$	$3s^2\ 3p^6$	0
Potassium	19	$1s^2\ 2s^2\ 2p^6\ 3s^2\ 3p^6\ 4s^1$	$4s^1$	+1
Calcium	20	$1s^2\ 2s^2\ 2p^6\ 3s^2\ 3p^6\ 4s^2$	$4s^2$	+2

For brevity, many chemists record the electron configuration of an atom by giving only its outermost subshell, like $4s^1$ for potassium or $4s^2$ for calcium. These electrons are most distant from the positive nucleus and, therefore, are most easily transferred between atoms in chemical reactions. These are the valence electrons.

For ions, the valence equals the electrical charge. In molecules, the various atoms are assigned chargelike values so the sum of the oxidation numbers equals the charge on the molecule. For example, in the H_2O molecule, each H has an oxidation number of +1, and the O is –2.

In Table 4-3, the common oxidation numbers in the last column are interpreted as the result of either losing the valence electrons (leaving a positive ion) or gaining enough electrons to fill that valence subshell. Table 4-4 compares three ions and a neutral atom.

Table 4-4 Electron Configurations of Ions

Chemical Element	*Valence Shell*	*Electron Transfer*	*Resulting Ion*	*Ion Configuration*
Cl	$3s^2 3p^5$	gain 1	Cl^-	$1s^2\ 2s^2\ 2p^6\ 3s^2\ 3p^6$
Ar	$3s^2 3p^6$	none	Ar^0	$1s^2\ 2s^2\ 2p^6\ 3s^2\ 3p^6$
K	$4s^1$	lose 1	K^+	$1s^2\ 2s^2\ 2p^6\ 3s^2\ 3p^6$
Ca	$4s^2$	lose 2	Ca^{2+}	$1s^2\ 2s^2\ 2p^6\ 3s^2\ 3p^6$

The charges on the chlorine, potassium, and calcium ions result from a strong tendency of valence electrons to adopt the stable configuration of the inert gases, with completely filled electronic shells. Notice that the three ions have electronic configurations identical to that of inert argon. These ions and the atom of argon are known as **isoelectronic.**

The Periodic Table

The pattern of elements in the periodic table reflects the progressive filling of electronic orbitals. The two columns on the left—the alkali metals and alkaline earths—show the addition of 1 and 2 electrons into *s*-type subshells. (See Figure 4-2.)

Figure 4-2 Filling of the *s* subshells.

1 H $1s^1$	2 He $1s^2$
3 Li $2s^1$	4 Be $2s^2$
11 Na $3s^1$	12 Mg $3s^2$
19 K $4s^1$	20 Ca $4s^2$

The loss of these *s*-subshell valence electrons explains the common +1 and +2 charges on ions of these elements, except for helium, which is chemically inert.

The six elements from boron through neon show the insertion of electrons into the lowest energy *p*-type subshell. (See Figure 4-3.)

Figure 4-3 Filling of the 2*p* subshell.

5	6	7	8	9	10
B	C	N	O	F	Ne
$2p^1$	$2p^2$	$2p^3$	$2p^4$	$2p^5$	$2p^6$

The same type of subshell is used to describe the electron configurations of elements in the underlying rows. (See Figure 4-4.)

Figure 4-4 Filling of the 3*p* subshell.

13	14	15	16	17	18
Al	Si	P	S	Cl	Ar
$3p^1$	$3p^2$	$3p^3$	$3p^4$	$3p^5$	$3p^6$

The three long rows of metallic elements in the middle of the periodic table, constituting the rectangle from scandium (21) to mercury (80), are the **transition metals.** Each of these three rows reflects the filling of a *d*-type subshell that holds up to 10 electrons. Figure 4-5 shows the valence subshell of the first series of transition metals. Notice the general increase in the number of electrons occupying the 3*d* subshell.

Figure 4-5 Filling of the 3*d* subshell.

21	22	23	24	25	26	27	28	29	30
Sc	Ti	V	Cr	Mn	Fe	Co	Ni	Cu	Zn
$3d^1$	$3d^2$	$3d^3$	$3d^5$	$3d^5$	$3d^6$	$3d^7$	$3d^8$	$3d^{10}$	$3d^{10}$

Anomalous configurations (Cr and Cu)

The anomalous electronic configuration of chromium and copper is interpreted as the displacement of 1 electron from an s orbital into a *d* orbital; these two elements have only one electron in the 4*s* subshell because the second electron was promoted into a 3*d* subshell. This example warns you that there are exceptions to the general pattern of electronic configurations of the elements. The complicated electronic structure of the transition metals is a consequence of the similar energy of various subshells, like the 4*s* and 3*d* subshells, which leads to multiple valence states for single elements. Vanadium, for example, shows valences of +2, +3, +4, or +5.

The two rows at the bottom of the periodic table are designated as the **lanthanides** and **actinides,** respectively. The lanthanides belong between elements 57 and 72, while the actinides belong between elements 89 and 104. (See Figure 4-6.)

Figure 4-6 The correct placement of the lanthanides and actinides in the periodic table.

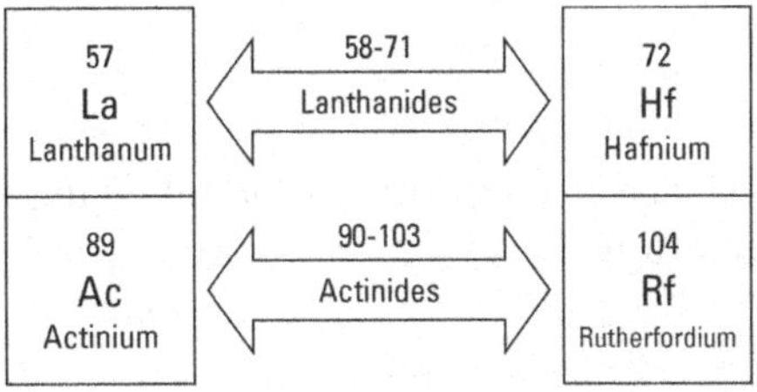

These two long rows of elements are traditionally moved to the base of the chart so the more important, lighter elements may be closer together for clarity. These two rows of metals each reflect the progressive addition of 14 electrons into an *f*-type subshell. The lanthanides occur in only trace amounts in nature and are often called **rare earths.** All the actinides have large, unstable nuclei that undergo spontaneous radioactive decay. Elements with atomic numbers 93 and higher were synthetically produced.

The outermost electrons of an atom generally determine the chemical behavior of that element. They determine the atom's size, charge, and ability to exchange electrons with other atoms. If you understand how the periodic table displays the pattern of electron configurations, you are on your way to mastering chemistry. You should know that rows of the periodic table show the filling of various subshells and that congeners in columns have similar electron subshells that are filled to the same degree.

Chapter Check-Out

1. What is the maximum number of electrons that can occupy one orbital?

a. 1
b. 2
c. 3

2. How many and what kind of subshells are in the $n = 2$ shell of an atom, and how many orbitals are present?

a. two subshells, *s* and *p*, and four orbitals, one *s*-orbital + three *p*-orbitals
b. one *p* subshell with three *p*-orbitals
c. two subshells, both *s* subshells with a total of two *s*-orbitals

3. Which is the ground state electronic configuration of neon (Ne), atomic number 10?

a. $1s^2\ 2s^4\ 2p^4$
b. $1s^2\ 1p^6\ 2s^2$
c. $1s^2\ 2s^2\ 2p^6$

4. How many valence electrons are in the following ground state electronic configuration? $1s^2\ 2s^2\ 2p^6\ 3s^1$

a. 7
b. 1
c. 3

5. Which element can be expected to form an ion with a +2 charge?

a. Na: $1s^2\ 2s^2\ 2p^6\ 3s^1$
b. O: $1s^2\ 2s^2\ 2p^4$
c. Mg: $1s^2\ 2s^2\ 2p^6\ 3s^2$

Answers: **1.** b **2.** a **3.** c **4.** b **5.** c

Chapter 5
CHEMICAL BONDING

Chapter Check-In

- ❑ Counting valence electrons
- ❑ Writing Lewis structures
- ❑ Forming covalent bonds
- ❑ Forming ionic bonds
- ❑ Sharing electrons unequally

One of the most important advances in chemistry during the twentieth century was the understanding of the way atoms bond (join together) to form compounds. Atoms either share pairs of electrons between them (a **covalent bond**), or they transfer one or more electrons from one atom to another to form positive and negative ions, which are held together because of their opposite charge (the **ionic bond**). The electrons farthest away from the nucleus, the **valence** electrons, are the ones involved in bonding. Lewis symbols for elements show only the valence electrons possessed by a particular element. The valence shell becomes uniquely stable with eight electrons. Many elements try to fill out the valence shell with eight electrons by forming bonds with other atoms.

Bonds between nonmetal elements are formed by sharing electrons between atoms, whereas the formation of bonds between metals and nonmetals involves the transfer of electrons forming ions of opposite charge. If an atom completely loses one or two electrons, it becomes an ion with a +1 or +2 charge, respectively, such as Na^+ or Ca^{2+}. If an atom gains one or two electrons, it forms an ion with a –1 or –2 charge, respectively, such as Cl^- or S^{2-}. Metals tend to lose electrons, whereas nonmetals tend to gain electrons in reactions. In all cases and no matter how it is done, the goal to achieve eight electrons in the outermost shell drives the formation of bonds between atoms.

Covalent Bonds

You may recall from the discussion of electron configurations (see Table 4-4) that a stable configuration is a completely filled *s*-type subshell and a *p*-type subshell. Only five elements have atoms with their valence *p*-subshells filled; these are the inert gases in the far right column of the periodic table. Their lack of chemical reactivity is explained by their stable electron configurations.

All other chemical elements need to lose or gain electrons to achieve electronic stability. Table 5-1 shows the stable electron configurations for the elements in the first three rows of the periodic table.

Table 5-1 Stable Inert Gas Electron Configurations

Element	*Atomic Number*	*Electron Configuration*	*Electron Capacity*	*Valence Shell*	*Valence Electrons*
He	2	$1s^2$	2	$1s^2$	2
Ne	10	$1s^2\ 2s^2\ 2p^6$	10	$2s^2\ 2p^6$	8
Ar	18	$1s^2\ 2s^2\ 2p^6\ 3s^2\ 3p^6$	18	$3s^2\ 3p^6$	8

Most atoms achieve a stable number of valence electrons by sharing electrons with other atoms. Begin with fluorine, element 9, which has the electron configuration $1s^2\ 2s^2\ 2p^5$. The orbitals of the valance electron shell are $2s\ 2p$, with two electrons in the $2s$ and five electrons in the $2p$. These seven valence electrons can be portrayed in a diagram devised by the American chemist Gilbert Lewis (1875–1946). In Figure 5-1, a Lewis diagram shows each valence electron as a single dot.

Figure 5-1 The valence electrons.

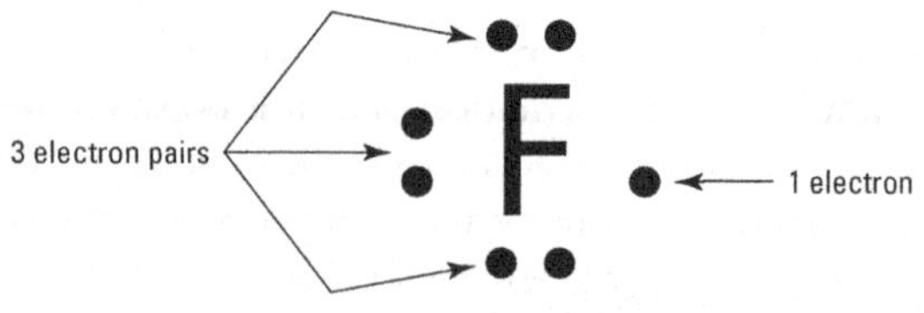

In Figure 5-2, two fluorine atoms can each fill their valence orbitals with eight electrons if they approach each other to share their single electrons.

Figure 5-2 Fluorine atoms sharing two electrons.

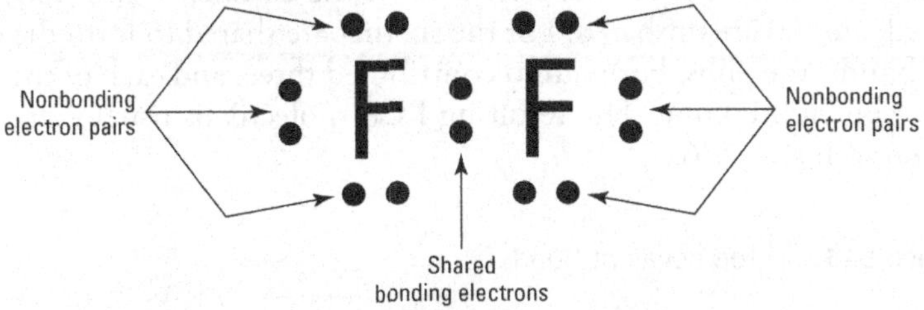

Count the electrons in the Lewis diagram; notice that there are 14 electrons, with each atom contributing 7. The two fluorine atoms form a stable F_2 molecule by sharing two electrons; this linkage is called a covalent bond.

You can determine the number of valence electrons for the light elements by counting the columns from the left. (See Figure 5-3.)

Figure 5-3 Counting valence electrons.

11	12	13	14	15	16	17
Na	Mg	Al	Si	P	S	Cl
$3s^1$	$3s^2$	$3s^2 3p^1$	$3s^2 3p^2$	$3s^2 3p^3$	$3s^2 3p^4$	$3s^2 3p^5$
1	2	3	4	5	6	7

Valence electrons increase to the right

Phosphorus has five valence electrons, and chlorine has seven, so their isolated atoms have Lewis configurations as shown in Figure 5-4.

Figure 5-4 The valence electrons of phosphorus and chlorine.

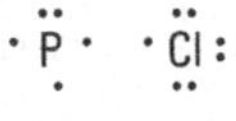

Phosphorus must combine with three chlorines to complete its valence shell. (See Figure 5-5.)

Figure 5-5 Six electrons shared between phosphorus and chlorine.

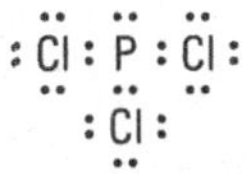

Study Figure 5-5 carefully. First, see that each atom is now surrounded by a full shell of eight valence electrons. Of the 26 valence electrons, 6 are shared, and 20 are unshared. For the six that are shared to form the covalent bonds, the phosphorus atom contributed three, and each of the chlorines contributed one. The resulting PCl_3 molecule is usually drawn as shown in Figure 5-6.

Figure 5-6 Three covalent bonds.

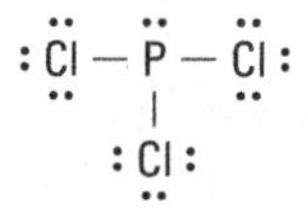

Each of the three lines represents the shared pair of electrons in a covalent bond. When lines are used to represent bonding pairs of electrons, the structure is often called a *structural formula.* Some textbooks omit the nonbonding electrons for simplicity.

Because the hydrogen atom has its single 1*s* orbital completed with only two electrons, the hydrogen chloride molecule is drawn as shown in Figure 5-7.

Figure 5-7 The hydrogen chloride molecule.

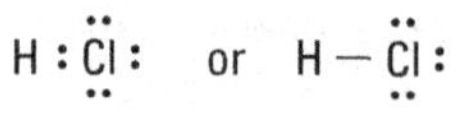

The hydrogen and chlorine atoms each donate one electron to the covalent bond. In the molecule, the hydrogen has completed its valence shell with two electrons, and the chlorine has a full shell with eight valence electrons.

In some molecules, bonded atoms share more than two electrons, as in ethylene (C_2H_4), where the two carbons share four electrons. (See Figure 5-8.)

Figure 5-8 A double bond between two carbon atoms.

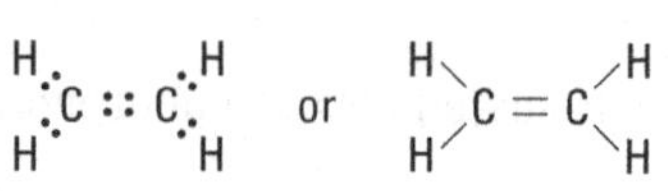

Notice that each carbon achieves eight electrons by this sharing. Because each shared pair constitutes a single covalent bond, the two shared pairs are called a *double bond.* The structure on the right side of Figure 5-8 shows this double bond of four shared electrons with two lines, and the left side of Figure 5-8 shows the double bond as two pairs of dots.

There are even triple bonds of six shared electrons, as in the nitrogen molecule. In N_2, each nitrogen atom contributes five valence electrons. Of the 10 electrons shown in Figure 5-9, four are nonbonding, and six comprise the triple bond holding the nitrogen atoms together.

Figure 5-9 A triple bond between two nitrogen atoms.

$$:N\vdots\vdots N: \quad \text{or} \quad :N\equiv N:$$

> **Problem 9:** Look at the periodic table and deduce the number of valence electrons for aluminum and oxygen from the positions of the columns for those two elements.
>
> **Problem 10:** Draw a Lewis diagram representing the electron configuration of the hydrogen sulfide molecule, H_2S.

Ionic Bonds

Although atoms with equal numbers of protons and electrons exhibit no electrical charge, it is common for atoms to attain the stable electronic configuration of the inert gases by either gaining or losing electrons. The metallic elements on the left side of the periodic table have electrons in excess of the stable configuration. Table 5-2 shows the electron loss necessary for three light metals to reach a stable electron structure.

Table 5-2 Metals Forming Stable Ions

Chemical Element	*Atomic Number*	*Total Electrons*	*Stable Number*	*Electron Transfer*	*Resulting Ion*
Neon	10	10	10	None	None
Sodium	11	11	10	Lose 1	Na^{+}
Magnesium	12	12	10	Lose 2	Mg^{2+}
Aluminum	13	13	10	Lose 3	Al^{3+}

The positive charge on the resulting metal ion is due to the atom possessing more nuclear protons than orbital electrons. The valence electrons are most distant from the nucleus; thus, they are weakly held by the electrostatic attraction of the protons and, consequently, are easily stripped from atoms of the metals.

By contrast, the nonmetallic elements on the right side of the periodic table have fewer electrons than that of a stable configuration and can most readily attain the stable configuration of the inert gases by gaining electrons. The negative charge on the resulting nonmetal ion is due to the atom possessing more orbital electrons than nuclear protons. Table 5-3 compares three nonmetals to the inert gas argon.

Table 5-3 Nonmetals Forming Stable Ions

Chemical Element	*Atomic Number*	*Total Electrons*	*Stable Number*	*Electron Transfer*	*Resulting Ion*
Phosphorus	15	15	18	Gain 3	P^{3-}
Sulfur	16	16	18	Gain 2	S^{2-}
Chlorine	17	17	18	Gain 1	Cl^{1-}
Argon	18	18	18	None	None

Because metallic elements tend to lose electrons and nonmetallic elements tend to gain electrons, a pair of contrasting elements will exchange electrons so that both achieve stable electronic configurations. The resulting ions of opposite charge have a strong force of electrostatic attraction, which is called an ionic bond. *Note:* This bond forms through the complete transfer of electrons from one atom to another, in contrast to the electron sharing of the covalent bond.

The force of attraction between two points of opposite electrical charge is given by Coulomb's law:

$$\text{force} = \frac{q_+ q_-}{d^2}$$

where q_+ is the positive charge, q_- is the negative charge, and d is the distance between the two charges. This law of electrostatic attraction can be used to measure the distance between two spherical ions because the charges can be considered to be located at the center of each sphere. (See Figure 5-10.)

Figure 5-10 The distance between ionic charges.

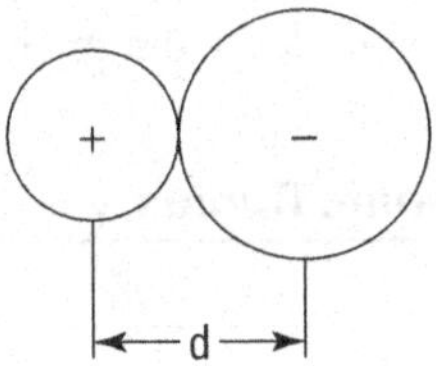

Notice that the distance between the centers of the two ions is the sum of radii of the ions. The appropriate electrostatic force then is calculated from the equation

$$\text{force} = \frac{q_C q_A}{(r_C + r_A)^2}$$

where q_C is the charge of the positive cation, q_A is the charge of the negative anion, and the denominator is the sum of their radii.

The strength of ionic bonding, therefore, depends on both the charges and the sizes of the two ions. Higher charges and smaller sizes produce stronger bonds. Table 5-4 shows the approximate radii of selected ions, which have the electronic configuration of an inert gas. The radii are in Å.

Table 5-4 Ionic Radii

Cations			***Anions***	
Li^+ 0.68	Be^{2+} 0.35	B^{3+} 0.23	O^{2-} 1.45	F^- 1.33
Na^+ 0.97	Mg^{2+} 0.66	Al^{3+} 0.51	S^{2-} 1.90	Cl^- 1.81
K^+ 1.33	Ca^{2+} 0.99	Sc^{3+} 0.73	Se^{2-} 2.02	Br^- 1.96
Rb^+ 1.47	Sr^{2+} 1.12	Y^{3+} 0.89	Te^{2-} 2.22	I^- 2.20
Cs^+ 1.67	Ba^{2+} 1.34	La^{3+} 1.06		

For ions of the same charge, the ionic radius increases as you go down any column because the elements of higher atomic number have a greater number of electrons in a series of electronic shells progressively farther from the nucleus. The change in ionic size along a row in the chart just above shows the effect of attraction by protons in the nucleus.

In Table 5-5, the five ions O^{2-} through Al^{3+} are all **isoelectronic**; that is, they have the same number of electrons in the same orbitals.

Table 5-5 Variation in Ionic Radius

Ion	O^{2-}	F^-	Ne^0	Na^+	Mg^{2+}	Al^{3+}
Nuclear protons	8	9	10	11	12	13
Orbital electrons	10	10	10	10	10	10
Ionic radius (Å)	1.45	1.33	1.10	0.97	0.66	0.51

For continuity, the neutral Ne atom is also in the chart, with its atomic radius. As you proceed to the right in Table 5-5, the greater number of protons attracts the electrons more strongly, producing progressively smaller ions.

Now use Coulomb's law to compare the strengths of the ionic bonds in crystals of magnesium oxide and lithium fluoride. The sizes of the four ions are taken from the tabulation of radii of cations and anions in Table 5-4.

$$\text{MgO:} \quad \frac{(2)(2)}{(0.66+1.45)^2} = 0.90$$

$$\text{LiF:} \quad \frac{(1)(1)}{(0.68+1.33)^2} = 0.25$$

Comparing the two relative forces of electrostatic attraction that you calculated, you can conclude that ionic bonding is considerably stronger in magnesium oxide. This affects the physical properties and chemical behavior of the two compounds. For example, the melting point of MgO (2,852°C) is much higher than that of LiF (845°C).

The strength of chemical bonding in various substances is commonly measured by the thermal energy (heat) needed to separate the bonded atoms or ions into individual atoms or ions.

Polar Bonds

Many substances contain bonds that are intermediate in character—between pure covalent and pure ionic bonds. Such **polar bonds** occur when one of the elements attracts the shared electrons more strongly than the other element. In hydrogen fluoride, for instance, the shared electrons are so much more attracted by fluorine than hydrogen that the sharing is unequal. (See Figure 5-11.)

Figure 5-11 Unequal sharing of electrons.

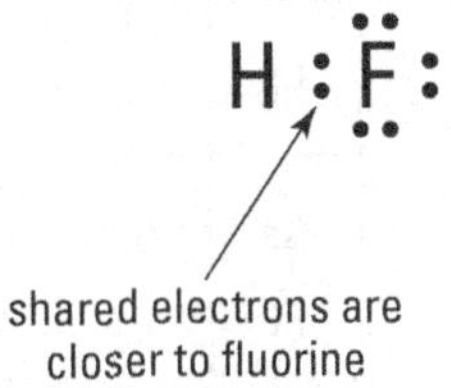

Due to the unequal sharing of the bonding electrons, the two atoms have fractional electrical charges, represented by the Greek letter delta (δ) in Figure 5-12.

Figure 5-12 Showing the partial charges of a polar bond.

$$\overset{\delta+}{\mathrm{H}} - \overset{\delta-}{\mathrm{F}}$$

Such an off-center or unequally shared covalent bond displays partial ionic character.

Around 1935, the American chemist Linus Pauling developed a scale of **electronegativity** to describe the attraction an element has for electrons in a chemical bond. The values in Figure 5-13 are higher for elements that more strongly attract electrons, which increases the likelihood of a negative partial charge on that atom.

Figure 5-13 The electronegativity of the elements.

H 2.1						
Li 1.0	Be 1.5	B 2.0	C 2.5	N 3.0	O 3.5	F 4.0
Na 0.9	Mg 1.2	Al 1.5	Si 1.8	P 2.1	S 2.5	Cl 3.0
K 0.8	Ca 1.0	Ga 1.6	Ge 1.8	As 2.0	Se 2.4	Br 2.8
Rb 0.8	Sr 1.0					I 2.4
Cs 0.7	Ba 0.9					

Many elements have been omitted to emphasize the basic pattern of electronegativity variation.

In Figure 5-13, you can see that the most electronegative element is fluorine. The nonmetals in the upper right corner have a strong tendency to gain electrons. The element of lowest electronegativity is cesium (Cs), in the lower-left corner. The relatively weak attraction for electrons by the alkali metals and alkaline earths is responsible for the loss of electrons by those elements.

Two atoms of the same electronegativity will share electrons equally in a pure covalent bond; therefore, any molecule that contains atoms of only one element, like H_2 or Cl_2, has pure covalent bonding. Two atoms of different electronegativities, however, will have either the unequally distributed electron distribution of a polar bond or the complete electron transfer of an ionic bond. Table 5-6 interprets the bonding between two elements as a function of the difference in their electronegativity.

Table 5-6 Electronegativity and Type of Bond

Electronegativity Difference	*Ionic Character (%)*	*Covalent Character (%)*	*Bond Type*
0.0	0	100	Covalent
0.5	5	95	Covalent
1.0	20	80	Covalent
1.5	40	60	Polar
2.0	60	40	Polar
2.5	75	25	Ionic
3.0	90	10	Ionic

Now use electronegativity to estimate the bond character in hydrogen sulfide, H_2S. The difference in electronegativities is

$$\underset{\text{S}}{2.5} - \underset{\text{H}}{2.1} = 0.4$$

You can interpolate this value in the first column of Table 5-6 to find that such a bond is about 4% ionic and 96% covalent, which is virtually a pure covalent bond.

Problem 11: Use the chart of electronegativity and the chart of bond types to interpret the bonding in magnesium chloride, $MgCl_2$.

Other Bonds

A polar bond between hydrogen and a very electronegative element, such as O, N, or F, allows a unique secondary bonding between the partially positive hydrogen and atoms with a partial negative charge. The attraction between hydrogen and these negative species is called the **hydrogen bond,** which is much weaker than the primary polar bond. Hydrogen bonding exists between water molecules because the electronegativity difference between hydrogen and oxygen is 1.4, indicating a polar bond of about 36% ionic character. Refer to Figure 5-14.

Figure 5-14 Polar bonds in the water molecule.

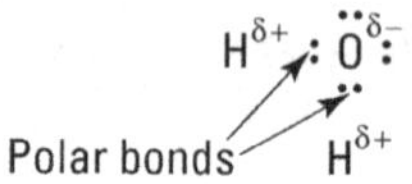

The polarity of the hydrogen-oxygen bond is responsible for much of the force of attraction between water molecules in liquid water. This strong force of attraction is responsible for the unusually high melting and **boiling points** of water. (See Figure 5-15.)

Figure 5-15 Hydrogen bonds between water molecules in liquid water.

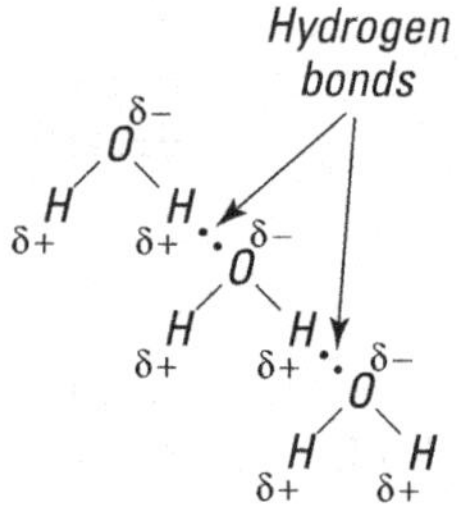

Chapter Check-Out

1. How many valence electrons are possessed by an oxygen atom?
   - a. 2
   - b. 4
   - c. 6

2. How many electrons are shared in a double bond between two atoms?
   - a. 2
   - b. 3
   - c. 4

3. When calcium (Ca) and chlorine (Cl) react, they each form an ion. What are the charges on the calcium ion and chloride ion, respectively?
   - a. +2 and –2
   - b. +1 and –2
   - c. +2 and –1

4. What is the formula of the compound formed between calcium and chlorine?
   - a. $CaCl_2$
   - b. Ca_2Cl
   - c. CaCl

5. What is the expected formula for the compound formed when one sulfur atom reacts with hydrogen?
   - a. HS
   - b. H_2S
   - c. H_6S

Answers: **1.** c **2.** c **3.** c **4.** a **5.** b

Chapter 6
ORGANIC COMPOUNDS

Chapter Check-In

- ❑ Learning why the chemistry of carbon is so extensive
- ❑ Understanding the importance of structure
- ❑ Seeing how carbon forms single, double, and triple bonds between atoms
- ❑ Forming derivatives of organic compounds

Organic chemistry is the chemistry of carbon, an element that forms strong chemical bonds to other carbon atoms as well as to many other elements like hydrogen, oxygen, nitrogen, and the halogens. Because of its versatility in forming covalent bonds, more than a million carbon compounds are known. Many are composed of only carbon and hydrogen, collectively called **hydrocarbons.** Most hydrocarbons are obtained from petroleum. Carbon always forms *four* covalent bonds (four shared pairs of electrons) that may be present as four single bonds per atom, or two single bonds and one double bond, or one single bond and one triple bond. With the ability of carbon to bond in different ways, an important part of organic chemistry concerns the structure of compounds. For example, three organic compounds have the identical molecular formula, C_5H_{12}, but they are different because each connects the five carbon atoms together in a different arrangement. Compounds with the same formula but different structures are called **isomers.** In organic chemistry, both structure and composition are of prime importance.

Organic chemistry is important because the vital biological molecules in living systems are largely organic compounds. Nearly all commonly used plastics are prepared from organic compounds.

Structural Formulas

The sixth element in the periodic table, carbon, has the electron configuration $1s^2\ 2s^2\ 2p^2$ and, thus, has four valence electrons in the unfilled orbitals of its second electron shell. To fill these orbitals to a stable set of eight valence electrons, a single carbon atom may share electrons with two, three, or even four other atoms. No other element forms such strong bonds to as many other atoms as carbon does. Moreover, multiple carbon atoms readily link together with single, double, or triple bonds. These factors make element number 6 unique in the entire periodic table. The number of carbon-based compounds is many times greater than the total of all compounds lacking carbon.

All types of life are based on carbon compounds, so the study of the chemistry of carbon is called **organic chemistry.** You should realize, however, that organic compounds are not necessarily derived from plants and animals. Hundreds of thousands of them have been *synthesized* (built) in the laboratory from simpler substances.

Figure 6-1 is an illustration of propane, one of the simplest organic compounds:

Figure 6-1 The structural formula of propane.

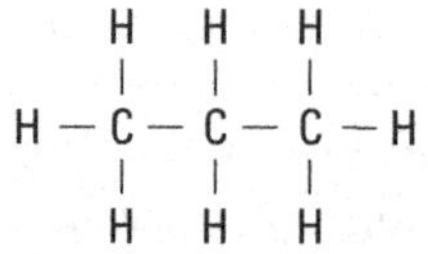

This representation is called a **structural formula,** in which lines depict two electron bonds between atoms. Look at the propane structure and observe that the four bonds to each carbon complete its valence orbitals with eight electrons.

In the diagram of propane, the most important feature is the chain of three carbons. Such carbon-carbon bonding is what generates the incredible variety of organic compounds. This linkage of carbon atoms can continue without limit. Just as propane has 3 bonded carbons, you can imagine organic compounds with 4 or 5 or 500 carbons in an extensive chain or network.

The structural formula for propane shows three axial carbon atoms and eight peripheral hydrogen atoms. The composition of propane can be more compactly expressed as C_3H_8. This representation is a **molecular**

formula. Such a formula does not directly tell how the various atoms are interbonded.

Compare two different compounds that have four linked carbon atoms. Refer to Figure 6-2.

Figure 6-2 Isomers of C_4H_{10}.

Structural formulas | Molecular formulas

Butane — C_4H_{10}

Isobutane — C_4H_{10}

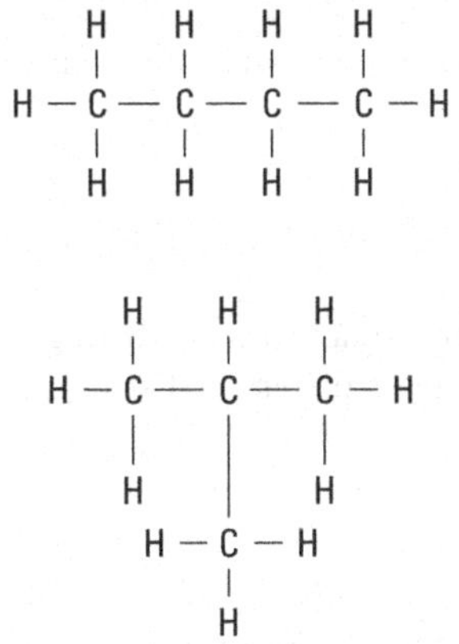

Although these two compounds have the same molecular formula (and, therefore, have identical chemical compositions), their structural formulas reveal a difference in the way that the four carbons are assembled. *Structure is just as essential as composition in organic chemistry.*

The two varieties of C_4H_{10} are called isomers, meaning that they have the same composition but differing structures. Structure affects both the physical properties and chemical reactivity of isomers. In the example of C_4H_{10} isomers, both exist as gases at room temperature, but they can be condensed easily to liquids by cooling or compression. The two liquids have different temperatures at which they boil. See Table 6-1.

Table 6-1 Structure and Boiling Point

Isomer	***Boiling Point***
Butane	–1°C
Isobutane	–12°C

The boiling behavior is consistent with their structures. The longest carbon chain in butane is four atoms, whereas the longest such chain in isobutane is only three atoms. The more compact molecules of isobutane escape from the liquid more readily, so the more volatile isobutane has a lower boiling point.

Chemists frequently write condensed structural formulas that omit the carbon-hydrogen bonds, as shown in Figure 6-3.

Figure 6-3 Condensed structural formulas.

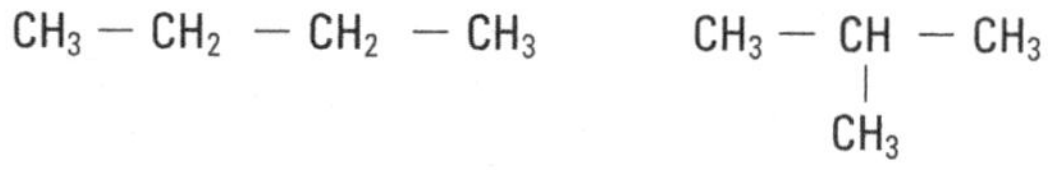

Butane   Isobutane

Notice that these condensed structural formulas still display the pattern of carbon-carbon bonding required to distinguish structural isomers.

Hydrocarbons

An infinite variety of compounds can be assembled from only carbon and hydrogen atoms. Such hydrocarbons are the simplest organic compounds, but they are also of prime economic importance because they include the constituents of petroleum and natural gas.

Propane, butane, and isobutane are all hydrocarbons with only single covalent bonds between carbon atoms. These hydrocarbons that lack double bonds, triple bonds, or ring structures make up the class called **alkanes.** See Table 6-2.

Table 6-2 The Six Simplest Alkanes

Compound Name	*Molecular Formula*	*Number of Isomers*
Methane	CH_4	1
Ethane	C_2H_6	1
Propane	C_3H_8	1
Butane	C_4H_{10}	2
Pentane	C_5H_{12}	3
Hexane	C_6H_{14}	5

As the number of carbon atoms increases, so does the number of ways that they can be connected to form different isomers. You should realize that isomers are defined by the pattern of bonding between the carbons.

The two molecules in Figure 6-4 are not different isomers; they are both butane. Despite the crooked carbon chain of the molecule on the right, it still has the same condensed structural formula, as shown in Figure 6-5.

Figure 6-4 Both molecules are the same isomer of butane.

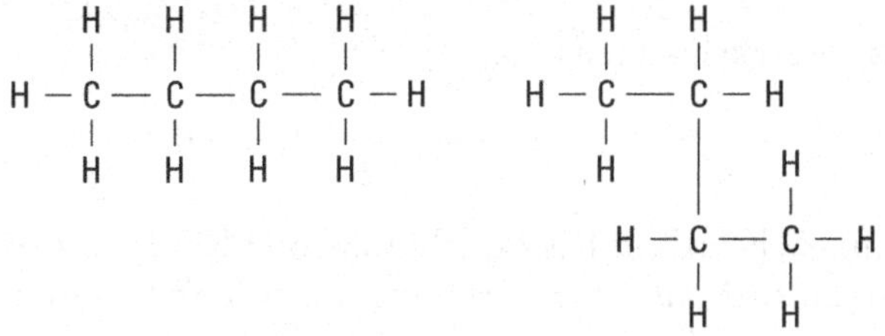

Figure 6-5 The condensed structural formula of butane.

$$CH_3 - CH_2 - CH_2 - CH_3$$

An **alkene** is a hydrocarbon with at least one double bond between carbons. The simplest alkene is ethylene, C_2H_4. See Figure 6-6.

Figure 6-6 Ethylene–an alkene.

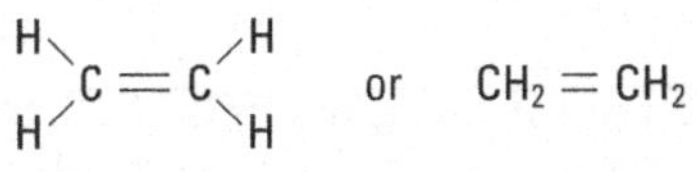

As is the case with the alkanes, each carbon atom in an alkene has precisely four bonds to fill its valence orbitals with eight electrons.

Another simple alkene is propene, C_3H_6. In Figure 6-7, propene demonstrates that alkenes can (and usually do) contain single bonds between some carbons. The existence of any double bond between carbons is the defining character.

Figure 6-7 Propene–one double and one single bond.

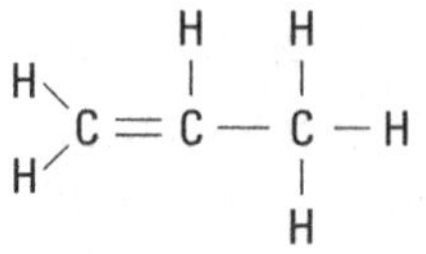 or $CH_2=CH-CH_3$

A hydrocarbon with a triple bond between carbons is an **alkyne,** and the simplest compound in this class is acetylene, C_2H_2, as shown in Figure 6-8.

Figure 6-8 Acetylene–an alkyne.

$$H-C\equiv C-H \quad \text{or} \quad CH\equiv CH$$

Once again, each carbon has exactly four bonds. Of course, the triple bond between carbons allows each carbon to bond to only one more atom. In acetylene, the single bond is to hydrogen, but in other alkynes, the single bond is to another carbon. Table 6-3 compares three hydrocarbons that contain the same number of carbon atoms.

Table 6-3 Compounds with Two Carbons

Hydrocarbon Class	***Compound Name***	***Molecular Formula***	***Carbon Bonding***
Alkane	Ethane	C_3H_6	Single
Alkene	Ethylene*	C_2H_4	Double
Alkyne	Acetylene**	C_2H_2	Triple

IUPAC retains these common, nonsystematic names: *Ethylene is the nonsystematic name for ethene; **acetylene is the nonsystematic name for ethyne. These common names are generally accepted by IUPAC.

Look at the third column of the chart and appreciate the diminishing hydrogen content of the compounds as the number of carbon-carbon bonds increases. Organic compounds with multiple carbon-carbon bonds readily react with hydrogen gas.

$$\underset{\substack{\text{ethylene}\\(\text{unsaturated})}}{C_2H_4} + \underset{\text{hydrogen}}{H_2} \rightarrow \underset{\substack{\text{ethane}\\(\text{saturated})}}{C_2H_6}$$

The hydrogenation reaction is possible only for compounds with double or triple bonds, and such compounds are said to be **unsaturated hydrocarbons.** The addition of the hydrogen to the carbon atoms that were double- or triple-bonded converts the unsaturated compound to a **saturated hydrocarbon** with only single bonds.

It is possible for long chains of carbons to loop around and form a closed ring structure. If you take the linear isomer of hexane in Figure 6-9 and delete the two hydrogens on the ends, the chain can form a hexagonal structure, as shown in Figure 6-10.

Figure 6-9 Hexane.

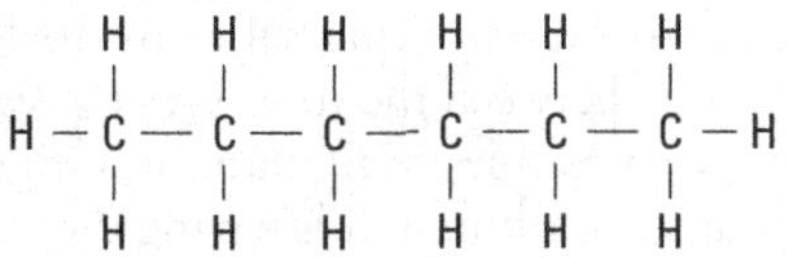

C_6H_{14}

Figure 6-10 Cyclohexane.

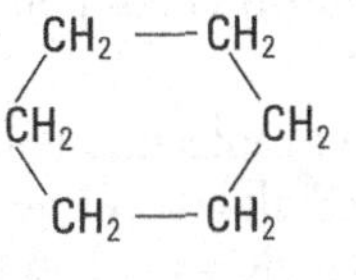

C_6H_{12}

Cyclohexane contains only single bonds and is representative of the simplest type of cyclic hydrocarbons.

A ring structure may possess double bonds, as in the following portrayal of the well-known hydrocarbon benzene, which has the composition C_6H_6. See Figure 6-11.

Figure 6-11 Benzene.

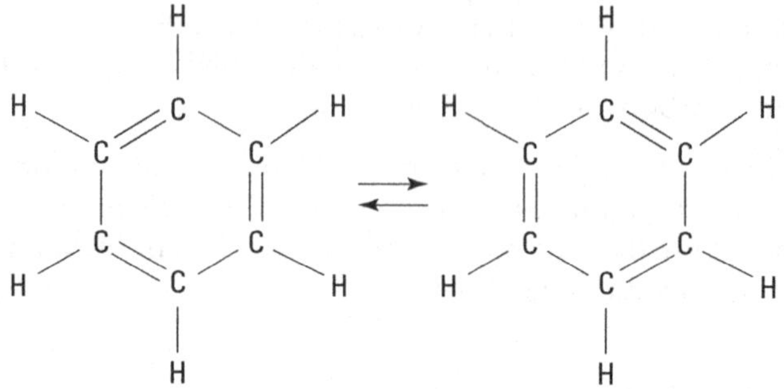

The two representations of the benzene ring differ in the location of the three double bonds. The arrows between the structures represent hypothetical transitions between the two possible configurations. Only one variety of benzene exists with all six carbon-carbon bonds having the same length and strength, so it seems best to regard the six extra electrons of the double bonds as being delocalized over the entire ring structure. Substances with benzene-like rings are called **aromatic** compounds.

Problem 12: Show the three isomers of pentane as condensed structural formulas.

Problem 13: Write a balanced molecular reaction for the hydrogenation of acetylene to a saturated alkane. How many liters of hydrogen gas are needed to react completely with 100 liters of acetylene?

Compounds with Additional Elements

The discussion of organic chemistry to this point has described only compounds of carbon and hydrogen. Although all organic compounds contain carbon, and almost all have hydrogen, most of them contain other elements as well. The most common other elements in organic compounds are oxygen, nitrogen, sulfur, and the halogens.

The halogens resemble hydrogen because they need to form a single covalent bond to achieve electronic stability. Consequently, a halogen atom may replace any hydrogen atom in a hydrocarbon. Figure 6-12 shows how fluorine or bromine atoms proxy for hydrogen in methane.

Figure 6-12 Methane and two derivatives.

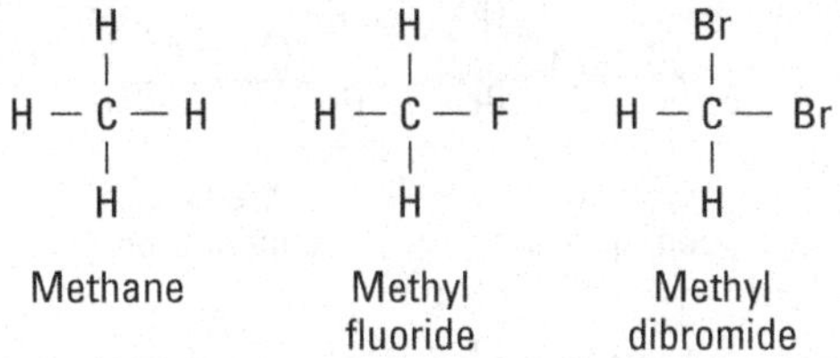

Halogens can replace any or all of the four hydrogens of methane. If the halogen is fluorine, the series of replacement compounds is

$$CH_4\ CH_3F\ CH_2F_2\ CHF_3\ CF_4$$

Such halogenated compounds are called *organic halides* or *alkyl halides.* The substituted atoms may be fluorine, chlorine, bromine, iodine, or any combination of these elements.

The previously mentioned ethylene molecule is *planar;* that is, all six atoms lie in a single plane because the double bond is rigid. In Figure 6-13, the stiff double bond prevents the molecule from being "twisted" around the axis between the carbon atoms.

Figure 6-13 Ethylene.

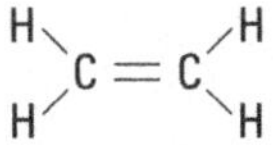

If a reaction substitutes a different atom such as a bromine atom for one or more hydrogen atoms, the resulting compound can exist in either of two different structural configurations. The configuration with the bromines adjacent is called *cis* (from the Latin derivative for "on this side"), whereas the configuration with bromines opposite is called *trans* (which means "on the other side"). The two configurations are different substances with unique chemical and physical properties. They are described as being *geometric isomers.* See Figure 6-14.

Figure 6-14 Geometric isomers.

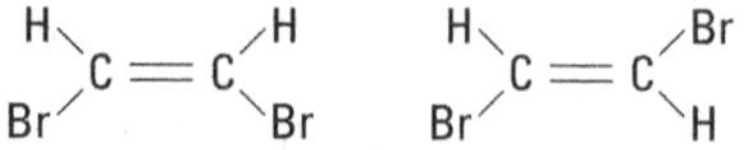

cis configuration   trans configuration

Figure 6-15 lists some common classes of organic compounds containing oxygen or nitrogen. The main carbon-bearing part of the compound attaches to the bond extending leftward in the second column. The examples use the ethyl C_2H_5– unit as the carbon chain attached to the functional group, but the immense number of organic compounds arises from the fact that virtually any carbon chain can be attached at that site.

Figure 6-15 Common functional groups.

Compound Class	Functional Group	Example of Group Attached to C_2H_5-
Alcohol	—O—H	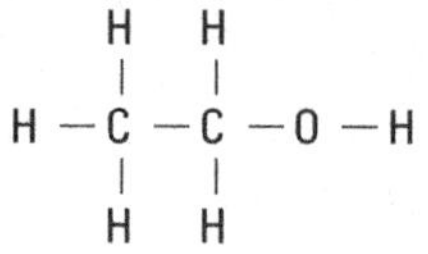
Aldehyde	—C(=O)—H	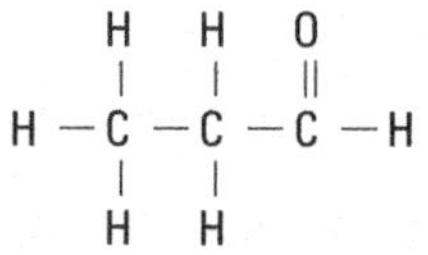
Carboxylic acid	—C(=O)—O—H	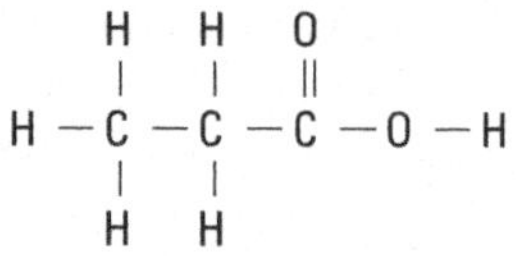
Amine	—N(—)—	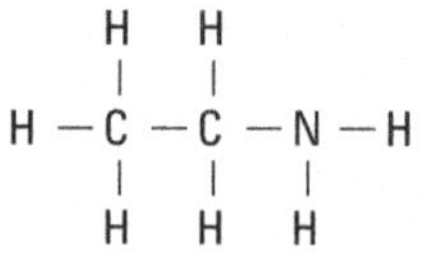

If you compare the carbon-oxygen bonding, you will observe that oxygens may be bonded to carbon by either single or double bonds.

Both alcohols and carboxylic acids have a single hydrogen bonded to an oxygen in the functional group. In aqueous solution, such hydrogens can become detached, producing slightly acidic solutions. More about this property is discussed in Chapter 9.

The amines contain nitrogen bonded to one, two, or three carbon chains. These compounds are derivatives of ammonia, hence the name of the class, as shown in Figure 6-16.

Figure 6-16 Ammonia.

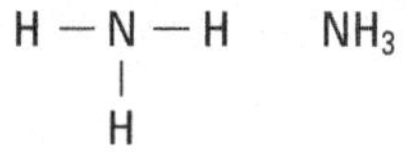

Consider three possible amines created by replacing hydrogen with the $-CH_3$ methyl group. See Figure 6-17.

Figure 6-17 Methyl derivatives of ammonia.

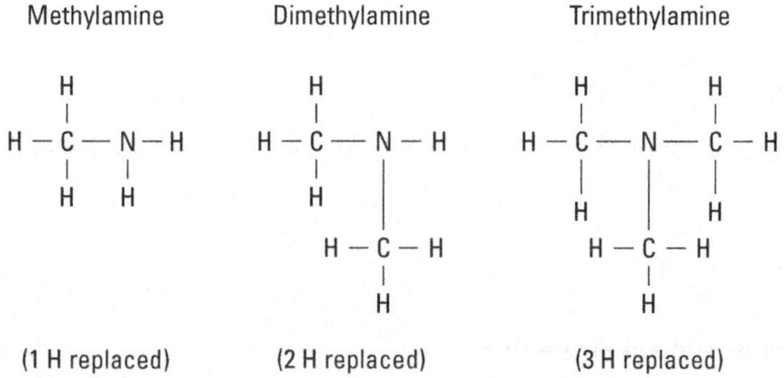

Of course, more complex carbon groups can be attached at any of the three bonds to nitrogen. Notice that the nitrogen atom is truly the core atom in an amine, in contrast to the functional groups in alcohols, aldehydes, and carboxylic acids, in each of which the functional group must be at the end of the molecule.

Problem 14: The oxidation of methyl alcohol produces a substance that has the composition of CH_2O. Draw the structure of this molecule and classify it on the basis of its functional group.

Chapter Check-Out

1. Substances that have the same composition but different structures are called ______.
   - **a.** alkenes
   - **b.** organic
   - **c.** isomers
2. A hydrocarbon with a triple bond between two carbon atoms is an ______.
   - **a.** alkane
   - **b.** alkene
   - **c.** alkyne
3. By reacting an alkene with hydrogen (H_2), the alkene will be changed to an ______.
   - **a.** alkane
   - **b.** alcohol
   - **c.** alkyne
4. Amines are derivatives of ______.
   - **a.** chlorine (Cl_2)
   - **b.** ammonia (NH_3)
   - **c.** water (H_2O)
5. The straight-chain hydrocarbon (not a ring) with a molecular formula of C_6H_{12} is an ______.
   - **a.** alkane
   - **b.** alkene
   - **c.** alkyne

Answers: **1.** c **2.** c **3.** a **4.** b **5.** b

Chapter 7
STATES OF MATTER

Chapter Check-In

- ❑ Learning the characteristics of the three states of matter: solid, liquid, and gas
- ❑ Understanding how pressure and temperature can determine and change the physical state of a substance

Nearly every substance can exist as a solid, a liquid, or a gas. These are the three common states of matter. Whether a substance is a solid, a liquid, or a gas depends on its temperature and the pressure placed on it. At room temperature (about 22°C) and at the normal pressure exerted by the atmosphere, water exists as a liquid, which can flow from one container to another. But if its temperature is lowered to –0.01°C, liquid water freezes to solid ice. Going the opposite direction in temperature and at this same pressure, water changes to a gas when the temperature exceeds 100°C. Changes in state can also occur by changing the pressure while holding temperature constant. The relationship between temperature and pressure and the three states of matter is easier to see when displayed in a phase diagram. Because phase diagrams provide so much information, they are known for thousands of substances.

Any change in phase is accompanied by the taking in or release of heat energy because, as change takes place, the attractive forces between molecules are being broken down or being formed. As solid water converts to liquid water, heat is absorbed as the forces between water molecules weaken, allowing the liquid to flow. The energy involved in phase changes is accurately known for many substances. The heat energy needed to warm or cool solids, liquids, and gases without changing phase is also accurately known.

Solids, Liquids, and Gases

The familiar compound H_2O provides the evidence that substances occur in three different physical classes called **states of matter.** At room temperature, H_2O is a dense fluid called a **liquid.** When this liquid is chilled to 0°C, it changes to a rigid **solid.** If the liquid is heated to 100°C, however, it abruptly expands to a tenuous fluid called vapor or **gas.**

Such different states of matter are not unique to H_2O. Almost all substances can exist in two or three of the fundamental states. Table 7-1 defines the states in terms of the shape and volume of substances. Because both liquids and gases flow readily, they are collectively referred to as **fluids.**

Table 7-1 Definitions of the States of Matter

State of Matter	*Shape of Substance*	*Volume of Substance*
Solid	Definite	Definite
Liquid	Indefinite	Definite
Gas	Indefinite	Indefinite

These states have different properties because they have distinct structures on the atomic or molecular scale. In a solid, the atoms are bonded strongly to the surrounding atoms so each is in a fixed position; if the solid structure has a regular pattern that is repeated throughout the solid, it is described as a **crystalline** structure. The atoms or molecules in a liquid are less strongly bonded to one another than in a solid of the same chemical composition, and consequently, they may shift their positions. The bonds between molecules in a liquid are, nevertheless, strong enough so that the molecules stay in contact with surrounding molecules. In a gas, the bonding between individual molecules is essentially zero, and individual molecules may move in all directions, allowing the vapor to expand throughout any container.

Phase Diagrams

Although the introductory example of H_2O mentioned changes of state caused by varying the temperature, it is known that variation of pressure can also produce such changes. In laboratory experiments, these two environmental factors—temperature and pressure—can each be varied or held constant; they are referred to as *independent variables.* Figure 7-1 assigns these variables to axes to form a plot that describes the physical condition at each point in the graph. The vertical axis is the pressure measured in atmospheres (atm).

Figure 7-1 The phase diagram for water.

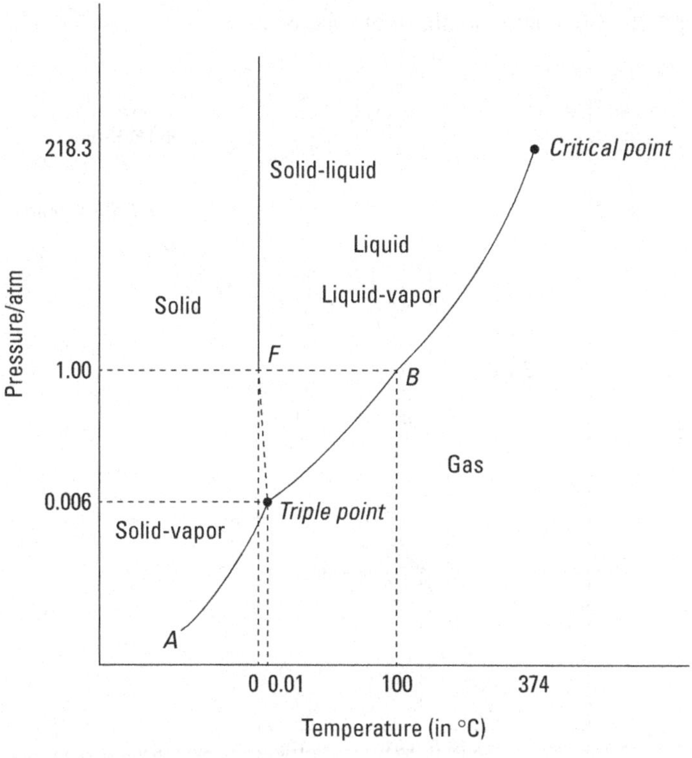

A temperature-pressure graph showing the various states of matter is a phase diagram. **Phase** refers to a single homogeneous physical state. Different phases have either different compositions or different physical states. In the preceding figure, there are three phases with the same composition in the solid, liquid, and gaseous states of matter.

Begin studying how both temperature and pressure determine the state of H_2O by taking some ice at a temperature of –10°C and pressure of 5 atmospheres, labeled *S* in Figure 7-2. If the pressure is held constant but the temperature is increased, the substance heats up along the dashed line marked *L*, melting to a liquid at point *m*, about –0.01°C. Alternatively, if you decrease the pressure on the initial solid *S*, while holding the temperature constant at –10°C, the conditions change downward along path G, and the ice vaporizes abruptly when the pressure has fallen to the point marked *n*, about 3×10^{-3} atm. Such a direct change from a solid to a gas is called **sublimation;** notice that there was no intervening liquid state.

Figure 7-2 Changing the phase of solid water.

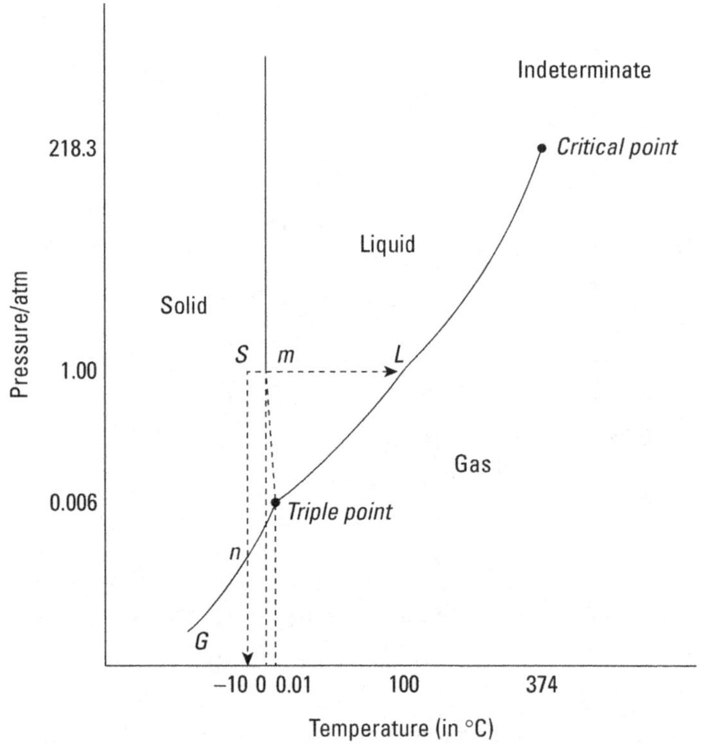

In the graph in Figure 7-3, study the possible state changes of an initial liquid marked *L*. The liquid is assumed to begin at 120°C and 30 atmospheres. The high pressure allows this liquid to exist at a temperature exceeding the 100°C boiling point at 1 atmosphere. If the pressure is maintained at a constant 30 atm, cooling the liquid *L* will produce a change to the left along path *S*, and the liquid will freeze at point *f* (about 0.01°C) to solid ice. A second course with constant pressure is heating *L*

toward G_1, and the liquid will abruptly vaporize at boiling point b_1 (about 235°C). Returning to the initial liquid L, you can imagine holding the temperature constant at 120°C while decreasing the pressure toward G_2. When the pressure falls to approximately 2 atm, the liquid will boil at point b_2. Boiling has been induced without heating the liquid.

Figure 7-3 Changing the phase of liquid water.

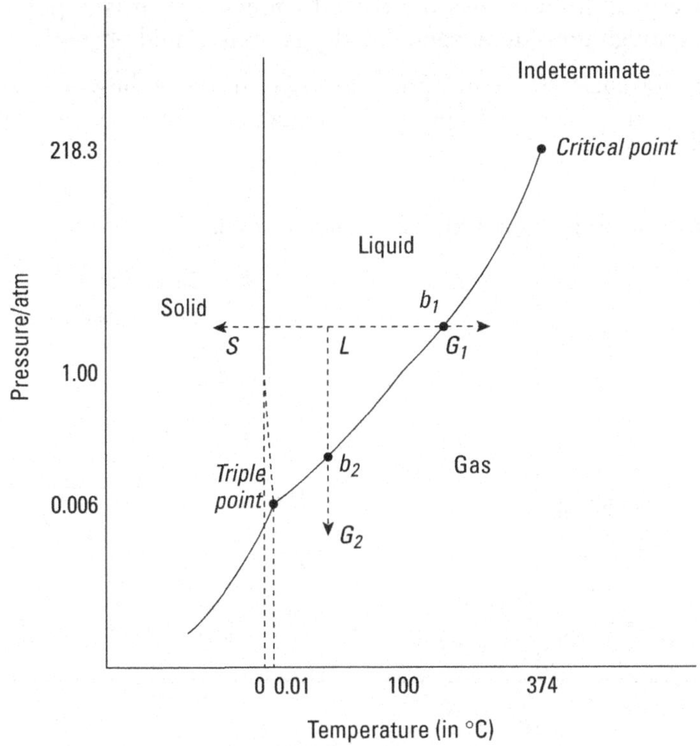

In summary, a change of state can be caused by varying only the temperature, varying only the pressure, or varying both temperature and pressure. Most random combinations of temperature and pressure fall within the three areas of a phase diagram in which only a single state is stable. The special temperature-pressure combinations plotted as lines in the phase diagram of H_2O (see Figure 7-2) are where two states can coexist. For example, both solid ice and liquid water are stable at precisely 0°C and 1 atm.

Look back at the large phase diagram (Figure 7-1) and notice the intersection of the three lines at 0.01°C and 6×10^{-3} atm. Only at this **triple point** can the solid, liquid, and vapor states of H_2O all coexist. The **critical point** is the highest temperature and highest pressure at which there is a difference between liquid and gas states. At either a temperature or a pressure over the critical point, only a single fluid state exists, and there is a smooth transition from a dense, liquid-like fluid to a tenuous, gas-like fluid. Beyond this point, the substance is often referred to as a super-critical fluid. Above the critical temperature, it is impossible to apply enough pressure to condense the gas to its liquid phase.

Each substance has its own phase diagram to display how temperature and pressure determine its properties. Figure 7-4 is the phase diagram for carbon dioxide.

Figure 7-4 The phase diagram for carbon dioxide.

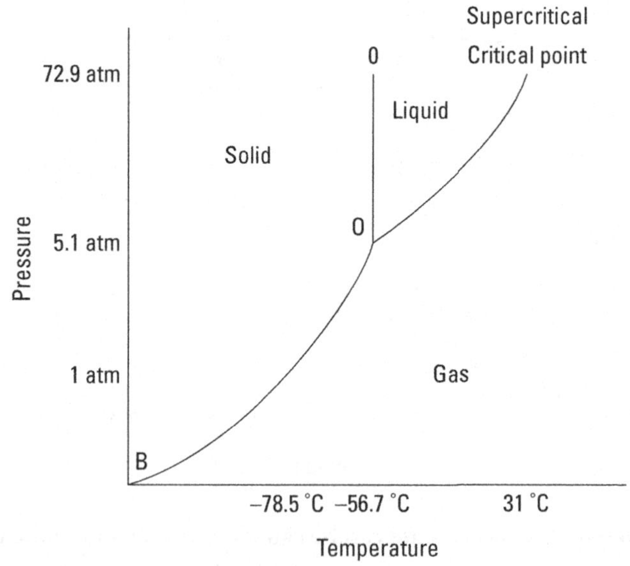

Use Figure 7-4 to answer the two practice problems.

Problem 15: What is the minimum pressure in atmospheres at which CO_2 can occur as a liquid?

Problem 16: If pressure is held at a uniform 3 atmospheres, at what temperature does solid CO_2 become unstable? What phase begins to appear at this temperature?

Heat Capacities and Transformations

For chemical reactions and phase transformations, the energy absorbed or liberated is measured as **heat.** The standard international unit for reporting heat is the **joule** (rhymes with school), which is defined as the energy needed to raise the temperature of 1 gram of water at 14.5°C by a single degree. The term *kilojoule* refers to 1,000 joules. Another unit of energy is the **calorie,** which is equal to 4.187 J. Conversely, a joule is 0.239 calories. The translation of calories to joules, or kilocalories to kilojoules, is so common in chemical calculations that you should memorize the conversion factors.

If a substance is heated without a change of state, the amount of heat required to change the temperature of 1 gram by 1°C is called the **specific heat capacity** of the substance. Similarly, the **molar heat capacity** is the amount of heat needed to raise the temperature of 1 mole of a substance by 1°C. Table 7-2 shows the heat capacities of several elements and compounds.

Table 7-2 Heat Capacities

Substance	*Joules per Degree per Gram*	*Joules per Degree per Mole*
$CaCO_3$	0.858	85.85
H_2O (liquid)	4.184	75.40
H_2O (solid)	2.029	36.57
MgO	0.208	35.08
Pb	0.130	26.87
Fe	0.452	25.24
Al	0.891	24.04

As an example of the use of the heat capacity values, calculate the joules required to heat 1 kilogram of aluminum from 10°C to 70°C. Multiply the grams of metal by the 60°C increase by the specific heat capacity:

$$1{,}000 \text{ grams} \times 60°\text{C} \times 0.891 \text{ cal/deg-g} = 53{,}472 \text{ joules}$$

It, therefore, requires 53.47 kilojoules of energy to heat this particular piece of aluminum. Conversely, if a kilogram of the same metal cooled from 70° to 10°C, 53.47 kJ of heat will be released into the environment.

You will realize that there is an abrupt change of energy when one state of matter is transformed into another. A considerable amount of energy is required to transform a low energy state to a higher energy state, like melting a solid to a liquid or vaporizing a liquid to a gas. The same quantity of energy is released upon the reverse transformation from a high energy state to a lower energy state, like condensing a gas to a liquid or freezing a liquid to a solid. Table 7-3 shows these energy values for H_2O.

Table 7-3 Heats of Transformation for H_2O

Change of State	*Associated Energy Term*	*Joules per Gram*	*Joules per Mole*	*Heat Flow*
Solid → liquid	Heat of fusion	333.9	6,008	Absorbed
Liquid → solid	Heat of crystallization	333.9	6,008	Released
Liquid → gas	Heat of vaporization	2256.8	40,648	Absorbed
Gas → liquid	Heat of condensation	2256.8	40,648	Released

Bear in mind that such transformations of state are *isothermal;* that is, they take place without any change in temperature of the substance. It takes 333.9 joules to change 1 gram of ice at 0°C to 1 gram of water at 0°C; the 333.9 joules are used to rearrange the molecules, which is done by overcoming intermolecular forces, from the crystalline order in the solid to the more irregular order in the liquid.

The data in the two previous tables permit some complex calculations of energy for changes of both state and temperature. Take a mole of water vapor at 100°C and cool it to ice at 0°. The energy released, which must be removed by the refrigeration process, comes from three distinct changes listed in Table 7-4.

Table 7-4 Example of Heat Calculation

Initial	*Final*	*Joules*	*Energy Source*
Vapor @ 100°C →	Liquid @ 100°C	40,648	Heat of condensation
Liquid @ 100°C →	Liquid @ 0°C	7,540	Heat capacity of water
Liquid @ 0°C →	Solid @ 0°C	6,008	Heat of crystallization
		54,195	Total heat released

You should make sure that you understand how each of the values in the third column is obtained. For example, the 7,540 joules is the molar heat capacity of water (75.40 j/deg) multiplied by the 100-degree change in temperature.

Notice especially that of the total heat released in this example, only 13.9% comes from lowering the temperature. Most of the heat comes from the two transformations of state—condensation and crystallization. For H_2O, the fact that the heat of condensation is almost seven times greater than the heat of crystallization may be interpreted as meaning that the molecular description of the liquid state is much more like the solid than the gas.

Problem 17: Use the data for H_2O in Table 7-2 and Table 7-3 to calculate the joules required to change 100 grams of ice at –40°C to water at 20°C.

Chapter Check-Out

1. The name of the process that converts a solid directly into a gas is called ______.
   a. boiling
   b. melting
   c. sublimation
2. Which phase of a substance has a definite volume but an indefinite shape?
   a. solid
   b. liquid
   c. gas
3. The point on a phase diagram at which the solid, liquid, and gaseous phases exist simultaneously is the ______.
   a. critical point
   b. triple point
   c. heat capacity
4. Which process absorbs heat?
   a. freezing a liquid to a solid
   b. converting gaseous water to liquid water
   c. vaporizing liquid water to gaseous water
5. If it requires 0.452 joules of heat to raise the temperature of 1.0 gram of iron 1.0°C, how many joules will be required to raise the temperature of 1.0 g of iron 15.0°C?
   a. 6.79 joules
   b. 62.76 joules
   c. 0.030 joules

Answers: **1.** c **2.** b **3.** b **4.** c **5.** a

Chapter 8
GASES

Chapter Check-In

- ❑ Learning how the volume of any gas is affected by temperature
- ❑ Using the Ideal Gas Equation
- ❑ Learning to use Avogadro's number

Of the three common states of matter, the gaseous state was most easily described by early scientists. As early as 1662, Robert Boyle showed how the volume of a gas, any gas, changed as the pressure applied to it was changed. Soon thereafter, the effects of temperature and the quantity of gas on the volume were discovered. The result of all these studies was a set of fundamental mathematical equations known as the gas laws that applied equally well to any gas, whether pure oxygen, nitrogen, or a mixture of the two. Through careful studies of gases reacting with one another, Amedeo Avogadro later concluded that equal volumes of different gases must contain the same number of molecules. For example, 10.0 L of oxygen contained the same number of oxygen molecules as there were nitrogen molecules in 10.0 L of nitrogen.

As time passed, it became clear that one mole of any gas contained the same number of molecules, 6.02×10^{23} molecules to be exact, a number known today as **Avogadro's number.** One mole of anything is Avogadro's number of that thing—atoms, molecules, people, or dollars—and that would be a lot of dollars!

Boyle's Law

Pressure is the amount of force exerted on one unit of area. The example of an ocean diver should make the concept clearer: The greater the depth the diver reaches, the greater the pressure due to the weight of the overlying water. Pressure is not unique to liquids but can be transmitted by gases and solids, too. At the surface of the Earth, the weight of the overlying air exerts a pressure equal to that generated by a column of mercury 760 mm high. The two most common units of pressure in chemical studies are atmosphere and millimeters of mercury.

$$1 \text{ atm} = 760 \text{ mm Hg}$$

However, the standard international unit for pressure is the Pascal.

$$1 \text{ atm} = 101325 \text{ Pa}$$

The English scientist Robert Boyle performed a series of experiments involving pressure and, in 1662, arrived at a general law—that the volume of a gas varies inversely with pressure.

$$PV = \text{constant}$$

This formulation has become established as **Boyle's law.** Of course, the relationship is valid *only if the temperature remains constant.*

As an example of the use of this law, consider an elastic balloon holding 5 L of air at the normal atmospheric pressure of 760 mm Hg. If an approaching storm causes the pressure to fall to 735 mm Hg, the balloon expands. The product of the initial pressure and volume is equal to the product of the final pressure and volume while the temperature is constant.

$$P_1 V_1 = P_2 V_2$$

$$(760 \text{ mm})(5 \text{ liters}) = (735 \text{ mm})(x \text{ liters})$$

$$x = \frac{(760)(5)}{735} = 5.17 \text{ liters}$$

It is important that you realize that pressure and volume vary inversely; therefore, an increase in either one necessitates a proportional decrease in the other.

Problem 18: Convert a pressure of 611 mm Hg to atmospheres.

Problem 19: If a gas at 1.13 atm pressure occupies 732 milliliters, what pressure is needed to reduce the volume to 500 milliliters?

Charles' Law

In 1787, the French inventor Jacques Charles, while investigating the inflation of his man-carrying hydrogen balloon, discovered that the volume of a gas varied directly with temperature. This relation can be written as

$$\frac{V}{T} = \text{constant}$$

and is called **Charles' law**. For this law to be valid, *the pressure must be held* ***constant,*** *and the temperature must be expressed on the absolute temperature or Kelvin scale.*

Because the volume of a gas decreases with falling temperature, scientists realized that a natural zero-point for temperature could be defined as the temperature at which the volume of a gas theoretically becomes zero. At a temperature of absolute zero, the volume of an ideal gas would be zero. The absolute temperature scale was devised by the English physicist Kelvin, so temperatures on this scale are called *Kelvin* (*K*) temperatures. The relationship of the Kelvin scale to the common Celsius scale must be memorized by every chemistry student:

$$K = °\text{C} + 273.15$$

Therefore, at normal pressure, water freezes at 273.15 K (0°C), which is called the **freezing point,** and boils at 373.15 K (100°C). Room temperature is approximately 293 K (20°C). Both temperature scales are used in tables of chemical values, and many simple errors arise from not noticing which scale is presented.

Use Charles' law to calculate the final volume of a gas that occupies 400 ml at 20°C and is subsequently heated to 300°C. Begin by converting both temperatures to the absolute scale:

$$\text{T}_1 = 20°\text{C} = 293.15\ K$$

$$\text{T}_2 = 300°\text{C} = 573.15\ K$$

Then substitute them into the constant ratio of Charles' law:

$$\frac{V_1}{T_2} = \frac{V_2}{T_2}$$

$$\frac{400 \text{ mL}}{293.15K} = \frac{x \text{ mL}}{573.15K}$$

$$x = \frac{(400)(573.15)}{(293.15)} = 782 \text{ mL}$$

When using Charles' law, remember that volume and *Kelvin temperature* vary directly; therefore, an increase in either requires a proportional increase in the other.

> **Problem 20:** A gas occupying 660 ml at a laboratory temperature of 20°C was refrigerated until it shrank to 125 ml. What is the temperature in degrees Celsius of the chilled gas?

Avogadro's Law

The volume of a gas is determined not only by the pressure and volume but also by the quantity of gas. When the quantity is given in moles, the mathematical relation is

$$\frac{V}{n} = \text{constant}$$

where *n* represents the number of moles of the gas. This relationship is known as **Avogadro's law** because, in 1811, Amedeo Avogadro of Italy proposed that equal volumes of all gases contain the same number of molecules. His theory explained why the volumes of gases in reactions are in ratios of small integers, as in the combustion of hydrogen:

$$\underset{\substack{\text{hydrogen} \\ (2\text{ volumes})}}{2H_2(g)} + \underset{\substack{\text{oxygen} \\ (1\text{ volume})}}{O_2(g)} \rightarrow \underset{\substack{\text{water vapor} \\ (2\text{ volumes})}}{2H_2O(g)}$$

You should recall from the review of the mole unit that the reaction coefficients (2, 1, and 2 in the preceding example) can be interpreted as molecules, or as moles, or as volumes. This is because Avogadro's law holds for all gases. The number of molecules in 1 mole of a gas is known to be 6.02×10^{23}, a value called Avogadro's number. As an example of the use of this important number, calculate the mass in grams of a single oxygen atom:

$$1 \text{ mole } O_2 = 2(16.00) = 32.00 \text{ grams}$$

$$1 \text{ molecule of } O_2 = \frac{32.00 \text{ grams}}{6.02 \times 10^{23}} = 5.32 \times 10^{-23} \text{ grams}$$

$$1 \text{ atom of O} = \frac{5.32 \times 10^{-23} \text{ grams}}{2} = 2.66 \times 10^{-23} \text{ grams}$$

> **Problem 21:** How many hydrogen atoms are there in 10 grams of methane, CH_4?

Ideal Gas Equation

The relations known as Boyle's law, Charles' law, and Avogadro's law can be combined into an exceedingly useful formula called the **Ideal Gas Equation,**

$$PV = nRT$$

where R denotes the gas constant:

$$R = 0.082 \frac{\text{liter} - \text{atm}}{K - \text{mole}}$$

The temperature is, as always in gas equations, measured in Kelvin.

This formula is strictly valid only for ideal gases—those in which the molecules are far enough apart so that intermolecular forces can be neglected. At high pressures, such forces cause significant departure from the Ideal Gas Equation, and more complicated equations have been devised to treat such cases. The Ideal Gas Equation, however, gives useful results for most gases at pressures less than 100 atmospheres.

$$V = \frac{nRT}{P}$$

$$V = \frac{(1 \text{ mole})\left(\frac{0.082 \text{ L} - \text{atm}}{K - \text{mole}}\right)(273.15\ K)}{(1 \text{ atm})}$$

$$V = 22.40 \text{ liters}$$

This is the value stated in the carbon dioxide reaction; you were asked to memorize that 1 mole of any gas occupies 22.4 liters at STP.

You should be able to use the Ideal Gas Equation to determine any one of the four quantities—pressure, volume, moles, or temperature—if you are given values for the other three.

One important application is to deduce the molecular mass and formula for a gas. Assume that you know that the hydrocarbon propylene is, by mass, 85.6% carbon and 14.4% hydrogen. Then the atomic ratios of the compound are

$$C = \frac{85.6}{12.01} = \frac{7.13}{1}$$

$$H = \frac{14.4}{1.01}$$

$$\frac{H}{C} = \frac{14.26}{7.13} = \frac{2}{1}$$

Therefore, the propylene molecule is some integral multiple of CH_2: It can be CH_2 or C_2H_4 or C_3H_6 or a yet larger molecule. Measuring the volume of 10 grams of propylene at STP yields 5.322 liters, which you can use to calculate its molecular mass.

$$\frac{5.322 \text{ liters}}{22.40 \text{ liters/mole}} = 0.2376 \text{ mole}$$

$$\frac{10 \text{ grams}}{0.2376 \text{ mole}} = 42.09 \text{ grams/mole (the molecular mass)}$$

Because the atomic masses of 1 CH_2 unit added together is 14.03, the molecule contains three such units. Consequently, the molecular formula for propylene is C_3H_6.

Problem 22: What is the volume occupied by 1 kilogram of carbon monoxide at 700°C and 0.1 atm?

Problem 23: The ozone molecule contains only oxygen atoms. Determine the molecular formula of ozone given that 2.3 grams occupies 1,073 milliliters at standard temperature and pressure.

Chapter Check-Out

1. If temperature remains constant and the pressure on 1.0 L of air doubles, the volume of the gas ______.
   a. increases to 2.0 L
   b. stays the same
   c. decreases to 0.5 L

2. The number of atoms in one mole of iron is ______.
   a. Avogadro's number
   b. Ideal Gas Equation constant
   c. one

3. The values referred to as standard temperature and pressure (STP) are ______.
   a. 1 atmosphere of pressure and 25°C
   b. 1 atmosphere of pressure and 0°C
   c. 1 atmosphere of pressure and 100°C

4. Charles' law relates how the volume of a gas varies with ______.
   a. absolute temperature
   b. the number of moles of gas
   c. the pressure applied to the gas

5. At standard temperature and pressure, 2.0 moles of nitrogen gas (N_2) has a volume of ______.
   a. 22.40 L
   b. 11.20 L
   c. 44.80 L

Answers: **1.** c **2.** a **3.** b **4.** a **5.** c

Chapter 9
SOLUTIONS

Chapter Check-In

- ❑ Learning how to express the concentration of the solute in a solution
- ❑ Predicting whether a compound is soluble in water
- ❑ Learning how to express the solubility of a compound in water
- ❑ Predicting how the concentration of solute will change the freezing and boiling temperatures of solutions

Most chemical reactions occur in **solutions**. This is because a substance dissolved in a **solvent**, the **solute**, will be in its smallest state of subdivision, existing as individual molecules or ions that will increase their ability to react with other molecules or ions. Most chemistry in the body takes place in solution; in the absence of the solution, much of the chemistry of life would not take place. You are familiar with solutions that are liquid, like iced tea and seawater, but solutions can also be gases, like the atmosphere, or solids, like a gold ring, which is a mixture of silver dissolved in gold.

Solutions are mixtures composed of two or more substances in ratios that can change. Sugar dissolved in iced tea is a solution, but you can add more sugar if you like, and you still have a solution. By contrast, compounds are also composed of two or more substances (usually elements) but in a ratio that cannot vary. In water, there are 8 grams of oxygen for each gram of hydrogen. It won't be water if that ratio changes.

The amount of a substance dissolved in a given amount of solvent is the **concentration** of the solute, which can be expressed in terms of **molarity** or **molality.** If you know the molarity of a solution, you can determine the exact volume of the solution that contains a desired amount of the solute.

Solutions freeze at lower temperatures and boil at higher temperatures than the pure solvent itself. If you know the concentration of the substance dissolved in the solvent, you can calculate how much lower the solution will freeze or how much higher it will boil than the pure solvent itself. You apply this fact when you add antifreeze to the water in the radiator of a car to form a solution that will not freeze in winter.

Concentration Units

A solution is a mixture of two or more substances that is of the same composition throughout. The host substance is a solvent, and the dissolved substance is a solute. Although the most familiar solvents are liquids, like water or ethyl alcohol, the general concept of a solution includes solvents that are gases or even solids.

In a solution, the ratio of solvent to solute is not fixed, and it can vary over a wide range, unlike compounds that are composed of definite, fixed ratios of the elements that make them up.

Seawater is an example of a liquid solution with water as solvent because the dissolved sodium chloride, calcium carbonate, magnesium bromide, and other solutes are of varying concentrations. Carbonated soda water is another liquid solution, but in this case, the solute is a gas—carbon dioxide.

Air is considered to be a gaseous solution with the abundant nitrogen as the solvent and scarcer oxygen as the solute.

An example of a solid solution is electrum, the alloy of gold and silver. The ratio of the two metals is not fixed but can range from nearly pure gold to nearly pure silver. The dominant metal is deemed to be the solvent in which the minor metal is dissolved.

The abundance of a solute is its concentration, and this characteristic can be reported with an intimidating variety of terms that you must master because the accurate description of solutions is central to chemical theory and laboratory practice.

One way to measure the concentration is to measure the relative masses of the constituents, usually expressed as **mass percents**. Take, for example, an electrum ingot that was formed by melting 62 grams of gold and 800 grams of silver and then letting the material cool and solidify. The following is the composition of the ingot in mass percent:

$$\text{Mass percent} = \frac{\text{mass of element}}{\text{mass of sample}} \times 100\%$$

$$\text{Gold} = \frac{62\text{ g}}{862\text{ g}} \times 100\% = 0.072 \times 100\% = 7.2\% \text{ (solvent)}$$

$$\text{Silver} = \frac{800\text{ g}}{862\text{ g}} \times 100\% = 0.928 \times 100\% = 92.8\% \text{ (solvent)}$$

You know, however, that gold and silver have different atomic masses (see Figure 9-1) and that the preceding percents do not represent relative numbers of atoms. You can calculate the number of moles of gold and silver in the electrum by dividing the masses of each metal by the corresponding atomic mass. Knowing the number of moles of each metal allows you to calculate the **mole fraction** of each element, that is, the fraction of atoms that are gold and the fraction that are silver.

Figure 9-1 Silver and gold.

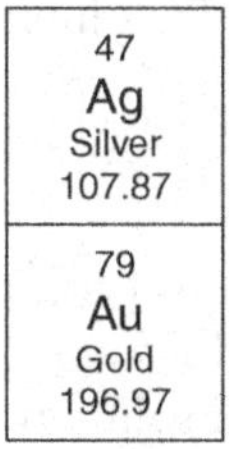

This is done by dividing the number of moles of one element by the sum of the number of moles of both gold and silver in the mixture using the following formula:

$$X_{\text{element}} = \frac{\text{moles of element}}{\text{total moles present in sample}}$$

$$\text{Moles of Ag} = \frac{800\text{ g}}{107.87\text{ g/mole}} = 7.416\text{ moles Ag}$$

$$\text{Moles of Au} = \frac{62\text{ g}}{196.97\text{ g/mole}} = 0.315\text{ moles Au}$$

$$X_{\text{Ag}} \frac{7.416\text{ mole}}{7.731\text{ mole}} = 0.959\text{ moles Ag}$$

$$X_{\text{Au}} = \frac{0.315\text{ mole}}{7.731\text{ mole}} = 0.041\text{ moles Au}$$

The mole fraction of silver is found to be 0.959. This means that 959 out of every 1,000 atoms in the mixture are silver atoms. The mole fraction of gold is 0.041, which means that 41 of every 1,000 atoms in the mixture are gold. The sum of the mole fractions of silver and gold will equal 1.00.

Although units of weight percent and mole fraction can be applied to all types of solutions, the most common concentration terms are molarity or molality. If water is the solvent, the solution is called an **aqueous** solution.

The molarity is the number of moles (or gram formula masses) of solute in 1 liter of *solution*. This unit is the most convenient one for laboratory work. A solution of calcium chloride that is 0.5 molar (abbreviated with an uppercase "M" as 0.5M) contains one-half mole of $CaCl_2$ (55.49 grams) in enough water to make the total volume 1 liter.

$$\text{Molarity }(\text{M}) = \frac{\text{moles of solute}}{\text{liters of solution}}$$

$$\text{M} = \frac{0.5 \text{ mole CaCl}_2}{1 \text{ liter H}_2\text{O}} = 0.5\text{M CaCl}_2$$

The other common unit for liquid solutions is *molality*, the number of moles of solute in 1 kilogram of *solvent*. Molality contrasts with molarity because it reports the amount of solute relative to the mass of the solvent, not the volume of the solution. A 2 molal solution of hydrogen fluoride, abbreviated 2 m (with a lowercase "m" for distinction from molarity), contains 2 moles of HF (40.02 grams) dissolved in 1,000 grams of H_2O.

$$\text{Molality }(\text{m}) = \frac{\text{moles of solute}}{\text{kilograms of solution}}$$

$$\text{m} = \frac{2.0 \text{ moles HF}}{1.0 \text{ kg H}_2\text{O}} = 2.0 \text{ m HF}$$

Molality is the preferred unit for certain types of calculations, although it is used less in laboratory work.

Problem 24: 80 grams of a simple sugar is added to 750 g of water. The sugar is glucose, with the composition $C_6H_{12}O_6$. What is the molality of glucose in the solution?

Problem 25: A solution is prepared by mixing 100 g of methyl alcohol (CH_3OH) and 100 g of water. What is the mole fraction of alcohol in the solution?

Solubility

Although some solutions, like one consisting of water and ethyl alcohol, can have any intermediate composition between the pure components, most solutions have an upper limit to the concentration of the solute. That limit is called the **solubility** of the substance. For example, in a liter of solution, the maximum amount of $CaSO_4$ dissolved is 0.667 grams, which is 0.0049 moles of that solute. Therefore, the solubility of calcium sulfate may be reported either as 0.667 grams per liter or as 0.0049 M.

A solution containing less solute than the maximum that can dissolve is known as a *dilute solution* and is said to be **unsaturated.** A solution containing as much solute as the solubility limit is described as saturated. Adding more of the solute to a saturated solution usually induces some of the solute to separate from the solution; if the separation is by means of the formation of crystals of the solute, these crystals are said to be precipitating from the solution. In some cases, a solution may contain more solute than the solubility limit. But, such a **supersaturated** state is unstable, and when precipitation begins, it will rapidly lower the concentration of the solute to the saturated level.

Table 9-1 is a useful summary of the relative solubilities of common chemical compounds. They are classified by their anions and listed from the most soluble at the top to the least soluble at the base. It is also helpful to remember that all compounds with the cation being an alkali metal or ammonium ion are highly soluble.

Table 9-1 Solubilities of Compounds

Class	*Anion*	*Description of Solubility*
Nitrates	NO_3^-	All are highly soluble.
Chlorates	ClO_3^-	All are highly soluble.
Chlorides	Cl^-	Highly soluble, except those of silver, lead(II) and mercury(I).
Bromides	Br^-	Highly soluble, except those of silver, lead(II) and mercury(I).
Iodides	I^-	Highly soluble, except those of silver, lead(II) and mercury(I).
Sulfates	SO_4^{2-}	Highly soluble, except those of strontium, barium, and lead(II).

(continued)

Table 9-1 ***(continued)***

Class	***Anion***	***Description of Solubility***
Sulfides	S^{2-}	Insoluble, except alkali metals, and ammonium.
Sulfites	SO_3^{2-}	Insoluble, except those of alkali metals and ammonium.
Hydroxides	OH^-	Insoluble, except those of alkali metals and ammonium.
Carbonates	CO_3^{2-}	Insoluble, except those of alkali metals and ammonium.
Phosphates	PO_4^{3-}	Insoluble, except those of alkali metals and ammonium.

It is important to realize that temperature markedly affects the solubility of most substances. For almost all salts, which are solid compounds composed of positive and negative ions (most often composed of both metallic and nonmetallic elements), an increase in temperature leads to an increase in the amount of the salt that will dissolve. Figure 9-2 shows the solubilities of potassium chloride (KCl) and potassium nitrate (KNO_3) as a function of temperature.

Figure 9-2 Temperature dependence of solubility.

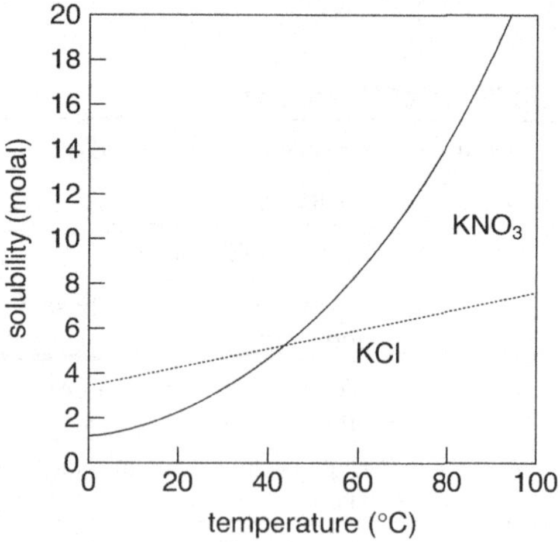

Notice that at 20°C, the KCl is more soluble, but at 40°C the KNO_3 has greater solubility. Although the solubilities of both salts increase with temperature, a given temperature rise enhances the solubility of the nitrate much more than the chloride.

When a compound containing ionic bonds is placed in water, the polar water molecules separate some or all of the substance into its cations and anions. The separation is referred to as ionic **dissociation**.

The concentrations of the ions may not equal the concentration of the solution. As an example, take 14.2 grams of sodium sulfate (Na_2SO_4) and add it to enough water to make 1 liter of solution. The sodium sulfate is highly soluble and dissolves completely, so the solution is 0.1 molar in Na_2SO_4.

$$\frac{14.2 \text{ g } Na_2SO_4/L}{142 \text{ g/mole}} = \frac{0.01 \text{ mole}}{L} = 0.1 \text{ M}$$

The salt, however, dissociates completely into ions:

$$Na_2SO_4(s) \rightarrow 2Na^+(aq) + SO_4^{2-}(aq)$$

In the preceding expression, the *(s)* denotes a solid and *(aq)* denotes an aqueous ion. In any reaction, the coefficients are proportional to the number of moles. So each mole of Na_2SO_4 yields two moles of Na^+ and one mole of SO_4^{2-}. The 0.1 M solution of Na_2SO_4 is, consequently, 0.2 M in Na^+ and 0.1 M in SO_4^{2-}.

For ionizing substances that are only slightly soluble, the concentrations of the ions multiply to a constant called the **solubility product** in a saturated solution. For a hypothetical compound CA, where the single cation is denoted by C and the anion by A, the solubility equation is

$$[C]\,[A] = K_{sp}$$

where the molar concentrations of the two ions are labeled with square brackets and the constant K_{sp} is the solubility product.

Many *binary compounds* (those with only two elements) contain more than one cation or anion. The general binary compound can be written C_xA_y, in which the subscripts mean the compound has *x* cations and *y* anions. In this case, the solubility equation is

$$[C]^x\,[A]^y = K_{sp}$$

Practice a solubility calculation using silver carbonate (Ag_2CO_3) as the solute. Dissocation of the salt yields three aqueous ions:

$$Ag_2CO_3(s) \rightarrow 2Ag^+(aq) + CO_3^{2-}(aq)$$

Table 9-2 states that K_{sp} for Ag_2CO_3 is 8.5×10^{-12}. The solubility equation involves the square of $[Ag^+]$ because each formula unit yields two ions of Ag^+.

$$[Ag^+]^2[CO_3^{2-}] = 8.5 \times 10^{-12}$$

Because the molarity of CO_3^{2-} is the same as the overall molarity of Ag_2CO_3 in the solution, call the carbonate concentration x and the silver ion concentration $2x$.

$$(2x)^2(x) = 8.5 \times 10^{-2}$$

$$4x^3 = 8.5 \times 10^{-12}$$

$$x^3 = 2.13 \times 10^{-12}$$

$$x = 1.29 \times 10^{-4} \text{ molar}$$

The solution, then, is 0.000129 M Ag_2CO_3, which is identical to the value found for the CO_3^{2-} concentration. Because the gram formula mass of Ag_2CO_3 is 275.75, each liter of solution contains

$$0.000129 \frac{\text{moles}}{\text{L}} \times 275.75 \frac{\text{g}}{\text{mole}} = 0.0355 \frac{\text{gram}}{\text{L}}$$

You can describe the solubility of silver carbonate as 0.0355 gram per liter.

Table 9-2 gives the solubility products for some important compounds that are sparingly soluble. The values are for 25°C, and each of them will vary with temperature.

Table 9-2 Solubility Products at 25°C

Compound	K_{sp}	*Compound*	K_{sp}
AgBr	5.4×10^{-13}	CuBr	6.3×10^{-9}
Ag_2CO_3	8.5×10^{-12}	CuCl	1.1×10^{-7}
$Al(OH)_3$	1.9×10^{-33}	$Cu(OH)_2$	1.6×10^{-19}
BaF_2	1.3×10^{-6}	$FeCO_3$	3.2×10^{-11}

Compound	K_{sp}	*Compound*	K_{sp}
$BaSO_4$	1.1×10^{-10}	$MgCO_3$	6.8×10^{-6}
CaF_2	1.5×10^{-10}	MgF_2	7.4×10^{-11}
$CaSO_4$	7.1×10^{-5}	$Mg(OH)_2$	5.6×10^{-12}

You need to use Table 9-2 to solve the following practice problems:

Problem 26: Suppose that you stirred 0.15 gram of cuprous chloride (CuCl) powder into a liter of water. Will the powder entirely dissolve?

Problem 27: Determine the solubility in grams per liter for aluminum hydroxide, $Al(OH)_3$.

Freezing and Boiling Points

For a solution with a liquid as solvent, the temperature at which it freezes to a solid is slightly lower than the **freezing point** of the pure solvent. This phenomenon is known as **freezing point depression** and is related in a simple manner to the concentration of the solute. The lowering of the freezing point is given by

$$\Delta T_1 = K_f m$$

where K_f is a constant that depends on the specific solvent and m is the molality of the molecules or ions solute. Table 9-3 gives data for several common solvents.

Table 9-3 Molal Freezing Point and Boiling Point Constants

Solvent	*Formula*	*Freezing Point (°C)*	K_f *(°C/molal)*	*Boiling Point (°C)*	K_b *(°C/molal)*
Water	H_2O	0.0	1.86	100.0	0.51
Acetic acid	CH_3COOH	17.0	3.90	118.1	3.07
Benzene	C_6H_6	5.5	4.90	80.2	2.53
Chloroform	$CHCl_3$	-63.5	4.68	61.2	3.63
Ethanol	C_2H_5OH	-114.7	1.99	78.4	1.22
Phenol	C_6H_5OH	43.0	7.40	181.0	3.56

Use the previous formula and the constant from Table 9-3 to calculate the temperature at which a solution of 50 grams of sucrose ($C_{12}H_{22}O_{11}$) in 400 grams of water will freeze. The molecular weight of sucrose is

$$12(12.01) + 22(1.01) + 11(16.00) = 342.34 \text{ g/mole}$$

so, the number of moles of sucrose is

$$\frac{50 \text{ grams}}{342.34 \text{ g/mole}} = 0.146 \text{ mole}$$

and the concentration of the solution in moles per kilogram of water is

$$\frac{0.146 \text{ moles}}{0.400 \text{ kg} H_2O} = 0.365 \text{ molal}$$

By taking the freezing point constant for water as 1.86 from Table 9-3 and then substituting the values into the equation for freezing point depression, you obtain the change in freezing temperature:

$$\Delta T_f = 1.86°\text{C/m} \times 0.365 \text{ m} = 0.68°\text{C}$$

Because the freezing point of pure water is 0°C, the sucrose solution freezes at –0.68°C.

A similar property of solutions is **boiling point elevation**. A solution boils at a slightly higher temperature than the pure solvent. The change in the boiling point is calculated from

$$\Delta T_b = K_b \text{m}$$

where K_b is the molal boiling point constant and m is the concentration of the solute expressed as molality. The boiling point data for some solvents are provided in Table 9-3.

Notice that the change in freezing or boiling temperature depends solely on the *nature of the solvent, not on the identity of the solute.*

One valuable use of these relationships is to determine the molecular mass of various dissolved substances. As an example, perform such a calculation to find the molecular mass of the organic compound santonic acid, which dissolves in benzene or chloroform. A solution of 50 grams of santonic acid in 300 grams of benzene boils at 81.91°C. Referring to Table 9-3 for the boiling point of pure benzene, the boiling point elevation is

$$81.91°\text{C} - 80.2°\text{C} = 1.71°\text{C} = \Delta T_b$$

Rearranging the boiling point equation to yield molality and substituting the molal boiling point constant from Table 9-3, you can derive the molality of the solution:

$$m = \frac{\Delta T_b}{K_b} = \frac{1.71^\circ\text{C}}{2.53^\circ\text{C}/m} = 0.676 \text{ molal}$$

That concentration is the number of moles per kilogram of benzene, but the solution used only 300 grams of the solvent. The moles of santonic acid is found as follows:

$$0.3 \text{ kg} \times 0.676 \text{ mole/kg} = 0.203 \text{ mole}$$

and the molecular weight is calculated as

$$\frac{50 \text{ grams}}{0.203 \text{ mole}} = 246.3 \text{ grams/mole}$$

The boiling point of a solution was used to determine that santonic acid has a molecular mass of approximately 246. You can also find this value by using the freezing point of the solution.

In the two previous examples, the sucrose and santonic acid existed in solution as molecules, instead of dissociating to ions. The latter case requires the total molality of all ionic species. Calculate the total ionic molality of a solution of 50.0 grams of aluminum bromide ($AlBr_3$) in 700 grams of water. Because the gram formula weight of $AlBr_3$ is

$$26.98 + 3(79.90) = 266.68 \text{ g/mole}$$

the amount of $AlBr_3$ in the solution is

$$\frac{50.0 \text{ grams}}{266.68 \text{ g/mole}} = 0.188 \text{ mole}$$

The concentration of the solution with respect to $AlBr_3$ formula units is

$$\frac{0.188 \text{ mole}}{0.700 \text{ kg solvent}} = 0.268 \text{ molal}$$

Each formula unit of the salt, however, yields one Al^{3+} and three Br^- ions:

$$AlBr_3\,(s) \rightarrow Al^{3+}\,(aq) + 3Br^-\,(aq)$$

So, the concentrations of the ions are

$$Al^{3+} = 0.268 \text{ molal}$$

$$Br^{-} = 3(0.268) = 0.804 \text{ molal}$$

$$Al^{3+} + Br^{-} = 1.072 \text{ molal}$$

The total concentration of ions is four times that of the salt. When calculating the change in freezing point or boiling point, the concentration of all the solute *particles* must be used, whether they are molecules or ions. The concentration of the ions in this solution of $AlBr_3$ is 1.072 molal, and this molality would be used to calculate ΔT_f and ΔT_b.

Problem 28: Calculate the boiling point of a solution of 10 grams of sodium chloride in 200 grams of water.

Problem 29: A solution of 100 grams of brucine in 1 kg chloroform freezes at –64.69°C. What is the molecular weight of brucine?

Chapter Check-Out

1. The molarity of a solution is the number of moles of solute in ______.

a. one kilogram of solvent
b. one liter of solution
c. one liter of solvent

2. Which compound is not expected to be very soluble in water?

a. NiS
b. $NaNO_3$
c. NH_4Cl

3. If the molarity of a solution of ammonium carbonate, $(NH_4)_2CO_3$, is 0.50 M, the molarity of ammonium ion, NH_4^+, is

a. 0.50 M
b. 1.0 M
c. 0.25 M

4. Which will have the lowest freezing temperature?

a. pure water
b. 0.5 M glucose in water
c. 1.0 M glucose in water

5. K_{sp} numbers can be used to ______.

a. determine the solubility of an ionic compound in water
b. determine the boiling point of a solution
c. determine the color of a solution

Answers: **1.** b **2.** a **3.** b **4.** c **5.** a

Chapter 10
ACIDS AND BASES

Chapter Check-In

- ❑ Using the pH scale to describe an acidic, neutral, or basic solution
- ❑ Learning the difference between strong acids and weak acids
- ❑ Learning what makes a solution basic and the compounds that can make it so
- ❑ Learning how water forms the ions that define an acid and base

Perhaps no two classes of compounds are more important in chemistry than acids and bases. All acids have several properties in common: They have a sour taste, and they all react with most metals to form hydrogen gas (H_2) and with baking soda to form carbon dioxide (CO_2). All acids turn blue **litmus** paper red, and their solutions conduct electricity because acids form ions when dissolved in water. All bases also share several common properties: They have a bitter taste; their solutions feel slippery like soapy water; and they turn red litmus paper blue (the opposite of acids). Solutions of bases also conduct electricity because they, too, form ions in water. Acids are similar because they produce a hydronium ion, H_3O^+ (*aq*), in water. Bases, on the other hand, all form a hydroxide ion, OH^-(*aq*), in water. These ions are responsible for the properties of acids and bases.

The pH scale was developed to express, in a convenient way, the concentration of hydrogen ion in solutions and is widely used when discussing acids and bases.

In the late 1800s, Svante Arrhenius defined an **acid** as a substance that increases the hydronium ion (H_3O^+) concentration in water, and a **base** as any substance that increases the hydroxide ion (OH^-) concentration in water. Acids and bases react with one another in a process called **neutralization** to form a **salt** and water. Hydrochloric acid neutralizes potassium hydroxide forming potassium chloride (a salt) and water:

$$HCl(aq) + KOH(aq) \rightarrow K^+(aq) + Cl^-(aq) + H_2O(l)$$

The pH Scale

Even distilled water contains some ions because a small fraction of water molecules dissociates to hydrogen and hydroxide ions:

$$\underset{\text{water molecule}}{H_2O(l)} \rightleftarrows \underset{\text{hydronium ion}}{H_3O^+(aq)} + \underset{\text{hydroxide ion}}{OH^-(aq)}$$

The double arrows indicate that the reaction proceeds either way. This condition of reciprocal reaction is called chemical **equilibrium**, and its importance to chemistry cannot be overemphasized. An equilibrium state is a stable, balanced condition, and it can be reproduced by many laboratory researchers. It also can be modeled well by simple mathematical equations.

The equilibrium between the water molecule and its ions shows that there are the same number of water molecules forming from union of the two ions as are dissociating into ions. The concentrations of the hydronium and hydroxide ions obey an equilibrium equation:

$$[H_3O^+]\,[OH^-] = K_w$$

where the concentrations are expressed in molarity and K_W is the *ion-product constant* for water. At a temperature of 25°C, the value of that constant is

$$K_W = 1.0 \times 10^{-14}$$

In pure water, the concentrations of H_3O^+ and OH^- must be equal because the dissociation of H_2O yields the same number of each of them. You can calculate the ionic concentrations from the equilibrium equation and the ion-product constant:

$$[H_3O^+][OH^-] = 1.0 \times 10^{-14}$$

$$[H_3O^+] = [OH^-] = x$$

$$x^2 = 1 \times 10^{-14}$$

$$x = 1.0 \times 10^{-7}\,\frac{\text{mole}}{\text{liter}} = [H_3O^+] = [OH^-]$$

The small value for the ion concentrations shows that in pure water only 1 in every 555,000,000 molecules has dissociated.

The substances classified as acids and bases affect the H_3O^+ and OH^- concentrations in solutions, but because the product of the two values must equal K_w, increasing one ion must depress the other. An acidic solution has more hydronium ions than hydroxide ions, whereas a basic—or **alkaline** solution has more hydroxide ions than hydronium ions. (See Table 10-1.)

Table 10-1 Ion Concentrations at 25°C

Solution	*[H^+]*	*[OH^-]*
Acidic	$>10^{-7}$	$<10^{-7}$
Neutral	10^{-7}	10^{-7}
Basic	$<10^{-7}$	$>10^{-7}$

The use of scientific notation to describe ion concentrations is somewhat cumbersome, and chemists have agreed to employ a **pH** scale to state the concentration of hydronium ions (the capital "H" in pH stands for hydrogen). They define pH to be the negative logarithm to base 10 of the hydronium ion concentration:

$$\text{pH} = -\log_{10}\,[H_3O^+\,(aq)]$$

Thus, the pH of neutral water at 25°C is calculated

$$\text{pH} = -\log(10^{-7}) = -(-7) = +7$$

and the pH of pure water is 7.

Calculate the pH of an acidic solution prepared by adding enough water to 5 grams of hydrochloric acid to make a solution that has a volume of 1 liter.

$$\frac{5\text{ g HCl/L}}{36.46\text{ g/mole}} = 0.137\text{ mole}$$

Assuming the hydrochloric acid is completely dissociated,

$$HCl(aq) \rightarrow H_3O^+(aq) + Cl^-(aq)$$

the concentration of the hydronium ions would be 0.137 M, and the pH of the solution equals

$$\text{pH} = -\log\,(0.137) = 0.86$$

Notice that this acidic pH is a positive number less than 7 because the calculation takes the *negative* logarithm of [H_3O^+]. Try this calculation now to make sure that you understand this point.

The pH of a solution may be estimated from color indicators that change hue with pH, like litmus or phenolphthalein papers; see Table 10-2. Where precise values are required, an electrical pH meter is utilized.

Table 10-2 pH Values at 25°C

Solution	*pH*	*Litmus*
Acidic	<7	Red
Neutral	7	Grey
Basic	>7	Blue

Problem 30: Calculate the pH of a solution with [OH^-] = 2.2×10^{-6} M, and classify the solution as acidic, neutral, or alkaline.

Strong and Weak Acids

Substances that dissociate completely into ions when placed in water are referred to as **strong electrolytes** because the high ionic concentration allows an electric current to pass through the solution. Most compounds with ionic bonds behave in this manner; sodium chloride is an example.

By contrast, other substances—like the simple sugar glucose—do not dissociate at all and exist in solution as molecules held together by strong covalent bonds. There also are substances—like sodium carbonate (Na_2CO_3)—that contain both ionic and covalent bonds. (See Figure 10-1.)

Figure 10-1 Ionic and covalent bonding in Na_2CO_3.

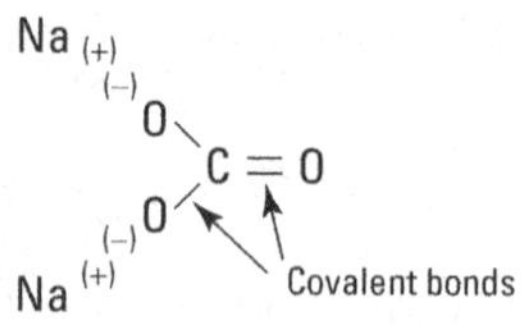

The sodium carbonate is a strong electrolyte, and each formula unit dissociates completely to form three ions when placed in water.

$$\underset{\text{sodium carbonate}}{Na_2CO_3(s)} \rightarrow \underset{\text{sodium cations}}{2Na^+(aq)} + \underset{\text{carbonate anions}}{CO_3^{2-}(aq)}$$

The carbonate anion is held intact by its internal covalent bonds.

Substances containing polar bonds of intermediate character commonly undergo only partial dissociation when placed in water; such substances are classed as **weak electrolytes**. An example is sulfurous acid:

$$H_2SO_3(aq) \rightleftarrows H_3O^+(aq) + HSO_3^-(aq)$$

A solution of sulfurous acid is dominated by molecules of H_2SO_3 with relatively scarce H_3O^+ and HSO_3^- ions. Make sure that you grasp the difference between this case and the previous example of the strong electrolyte Na_2CO_3, which completely dissociates into ions.

Acids and bases are usefully sorted into strong and weak classes, depending on their degree of ionization in aqueous solution.

The dissociation of any acid may be written as an equilibrium reaction:

$$HA(aq) \rightleftarrows H^+(aq) + A^-(aq)$$

where A denotes the anion of the particular acid. The concentrations of the three solute species are related by the equilibrium equation

$$\frac{[H^+][A^-]}{[HA]} = K_a$$

where K_a is the **acid ionization constant** (or merely acid constant). Different acids have different K_a values—the higher the value, the greater the degree of ionization of the acid in solution. Strong acids, therefore, have larger K_a than do weak acids.

Table 10-3 gives acid ionization constants for several familiar acids at 25°C. The values for the strong acids are not well defined; therefore, the values are stated only in orders of magnitude. Examine the "Ions" column and see how every acid yields a hydronium ion and a complementary anion in solution.

Table 10-3 Some Common Acids

Acid	*Formula*	*Ions*	K_a
Hydrochloric	HCl	H^+, Cl^-	10^7
Chloric	$HClO_3$	H^+, ClO_3^-	10^3
Sulfuric	H_2SO_4	H^+, HSO_4^-	10^2
Nitric	HNO_3	H^+, NO_3^-	10^1
Sulfurous	H_2SO_3	H^+, HSO_3^-	1.5×10^{-2}
Phosphoric	H_3PO_4	H^+, $H_2PO_4^-$	7.5×10^{-3}
Acetic	CH_3COOH	H^+, CH_3COO^-	1.8×10^{-5}
Carbonic	H_2CO_3	H^+, HCO_3^-	4.3×10^{-7}

Use the equilibrium equation and data from the preceding chart to calculate the concentrations of solutes in a 1 M solution of carbonic acid. The unknown concentrations of the three species may be written

$$\underset{1-x}{H_2CO_3(aq)} \rightleftarrows \underset{x}{H_3O^+(aq)} + \underset{x}{HCO_3^-(aq)}$$

where x represents the amount of H_2CO_3 that has dissociated to the pair of ions. Substituting these algebraic values into the equilibrium equation,

$$\frac{[H_3O^+][HCO_3^-]}{[H_2CO_3]} = \frac{x^2}{1-x} = K_a = 4.3 \times 10^{-7}$$

To solve the quadratic equation by approximation, assume that x is so much less than 1 (carbonic acid is weak and only slightly ionized) that the denominator $1 - x$ may be approximated by 1, yielding the much simpler equation

$$x^2 = 4.3 \times 10^{-7}$$

$$x = 6.56 \times 10^{-4} = [H_3O^+]$$

This H_3O^+ concentration is, as conjectured, much less than the nearly 1 molarity of the H_2CO_3, so the approximation is valid. A hydronium ion concentration of 6.56×10^{-4} corresponds to a pH of 3.18.

You will recall from the review of organic chemistry that carboxylic acids have a single hydrogen bonded to an oxygen in the functional group. (See Figure 10-2.) To a very small extent, this hydrogen can dissociate in an aqueous solution. Therefore, members of this class of organic compounds are weak acids.

Figure 10-2 Carboxylic acids.

Class	Compound	Formula	K_a
Carboxylic acids	acetic acid	$CH_3-C(=O)-O-H$	1.8×10^{-5}
	propionic acid	$CH_3CH_2-C(=O)-O-H$	1.3×10^{-5}

Summarize the treatment of acids so far. A strong acid is virtually completely dissociated in aqueous solution, so the H_3O^+ concentration is essentially identical to the concentration of the solution—for a 0.5 M solution of HCl, $[H_3O^+] = 0.5$ M. But because weak acids are only slightly dissociated, the concentrations of the ions in such acids must be calculated using the appropriate acid constant.

> **Problem 31:** If an aqueous solution of acetic acid is to have a pH of 3, how many moles of acetic acid are needed to prepare 1 liter of the solution?

Two Types of Bases

For bases, the concentration of OH^- must exceed the concentration of H_3O^+ in the solution. This imbalance can be created in two different ways.

First, the base can be a hydroxide, which merely dissociates to yield hydroxide ions:

$$MOH(aq) \rightleftarrows M^+(aq) + OH^-(aq)$$

where M represents the cation, usually a metal. The most familiar bases are such hydroxides. (See Table 10-4.)

Table 10-4 Common Bases

Base	*Formula*	*Ions*
Sodium hydroxide	NaOH	Na^+ OH^-
Potassium hydroxide	KOH	K^+ OH^-
Calcium hydroxide	$Ca(OH)_2$	Ca^{2+} $2OH^-$
Aqueous ammonia	NH_3 (*aq*)	NH_4^+ OH^-

The second type of base acts by extracting a hydrogen ion from a water molecule, leaving a hydroxide ion:

$$\underset{\text{base}}{B(aq)} + \underset{\text{water}}{H_2O} \rightleftarrows \underset{\text{acid}}{BH^+(aq)} + \underset{\text{hydroxide ion}}{OH^-(aq)}$$

An example of this second type of base that is not a hydroxide can be an ammonia molecule in water (aqueous ammonia):

$$\underset{\substack{\text{ammonia}\\\text{(base)}}}{NH_3(aq)} + \underset{\text{water}}{H_2O} \rightleftarrows \underset{\substack{\text{ammonium ion}\\\text{(acid)}}}{NH_4^+(aq)} + \underset{\text{hydroxide ion}}{OH^-(aq)}$$

Ammonia acts as a base by stripping a proton from a water molecule, leaving an increased OH^- concentration. Notice in the equilibrium reaction that NH_4^+ and NH_3 are a **conjugate** acid-base pair, related by transferring a single proton. Similarly, water acts as an acid by donating a proton to ammonia. H_2O and OH^- are a conjugate acid-base pair, related by the loss of a single proton.

Alternatively, the base may be a particular kind of negative ion with a high attraction for a hydrogen ion:

$$B^-(aq) + H_2O \rightleftarrows BH(aq) + OH^-(aq)$$

In 1923, the English chemist Thomas Lowry and the Danish chemist Johannes Brønsted defined an acid and base in another way. An acid is a substance that can donate a proton, and a base is a substance that can accept a proton.

> **Problem 32:** The bicarbonate ion HCO_3^- may serve as either a Brønsted-Lowry acid or base. When it acts as an acid, what is its conjugate base? When it behaves as a base, what is its conjugate acid?

Polyprotic Acids

Many acids contain two or more ionizable hydrogens. There are two in carbonic acid, H_2CO_3, and three in phosphoric acid, H_3PO_4. For any such multiple hydrogen acid, the first hydrogen is most easily removed, and the last hydrogen is removed with the greatest difficulty. These acids are called **polyprotic** (many protons) acids. The multiple acid ionization constants for each acid measure the degree of dissociation of the successive hydrogens.

Table 10-5 gives ionization data for four series of polyprotic acids. The integer in parentheses after the name denotes which hydrogen is being ionized, where (1) is the first and most easily ionized hydrogen.

Table 10-5 Four Series of Polyprotic Acids

Acid	*Formula*	*Conjugate Base*	K_a
Sulfuric (1)	H_2SO_4	HSO_4^-	About 10^{+2}
Sulfuric (2)	HSO_4^-	SO_4^{2-}	1.2×10^{-2}
Sulfurous (1)	H_2SO_3	HSO_3^-	1.6×10^{-2}
Sulfurous (2)	HSO_3^-	SO_3^{2-}	8.3×10^{-8}
Phosphoric (1)	H_3PO_4	$H_2PO_4^-$	7.5×10^{-3}
Phosphoric (2)	$H_2PO_4^{2-}$	HPO_4^{2-}	6.2×10^{-8}
Phosphoric (3)	HPO_4^{2-}	PO_4^{3-}	3.2×10^{-13}
Carbonic (1)	H_2CO_3	HCO_3^{3-}	4.3×10^{-7}
Carbonic (2)	HCO_3^-	CO_3^{2-}	5.2×10^{-11}

Remember: The strongest acids dissociate most readily. Of the nine acids listed in Table 10-5, the strongest is sulfuric (1), with the highest acid ionization constant, and the weakest is phosphoric (3).

Here are the chemical equations for the three successive ionizations of phosphoric acid:

First dissociation $H_3PO_4(aq) \rightleftarrows H_3O^+(aq) + H_2PO_4^-(aq)$

Second dissociation $H_3PO_4^-(aq) \rightleftarrows H_3O^+(aq) HPO_4^{2-}(aq)$

Third dissociation $HPO_4^{2-}(aq) \rightleftarrows H_3O^+(aq) + PO_4^{3-}(aq)$

Consequently, an aqueous solution of phosphoric acid contains all the following molecules and ions in various concentrations:

$$H_3PO_4 \quad H_2O \quad H_3O^+ \quad OH^- \quad H_2PO_4^- \quad HPO_4^{2-} \quad PO_4^{3-}$$

Consulting the table of the dissociation constants K_a's for phosphoric acid shows that the first dissociation is much greater than the second, about 100,000 times greater. This means nearly all the H_3O^+ (*aq*) in the solution comes from the first step of dissociation. The second and third steps add very little H_3O^+ (*aq*) to the solution. So a solution of phosphoric acid will contain H_3PO_4 molecules in highest concentration with smaller, and nearly equal, concentrations of H_3O^+ and $H_2PO_4^-$. The HPO_4^{2-} and PO_4^{3-} ions are present in very small concentrations.

Problem 33: What is the principle species in a solution of sulfurous acid, H_2SO_3, a weak polyprotic acid? List H_2SO_3, HSO_3^-, SO_3^{2-} and H^+ in order of decreasing concentration.

Chapter Check-Out

1. A solution has a hydronium ion concentration of 1×10^{-4} M. What is the pH of the solution, and is it considered an acidic, neutral, or basic solution?
   - **a.** 7, neutral
   - **b.** 10, basic
   - **c.** 4, acidic
2. Which compound is an acid in water?
   - **a.** NaOH
   - **b.** HCl
   - **c.** NH_3
3. Which of the following is a weak acid?
   - **a.** HCl
   - **b.** H_2SO_4
   - **c.** H_2CO_3
4. What is the molarity of hydronium ion in a 0.0050 M solution of HNO_3?
   - **a.** 0.0050 M
   - **b.** less than 0.0050 M
   - **c.** more than 0.0050 M
5. What is the pH of a solution made by dissolving 0.040 mole of HCl in 3500 mL of solution?
   - **a.** 1.94
   - **b.** 1.40
   - **c.** 0.76

Answers: **1.** c **2.** b **3.** c **4.** a **5.** a

Chapter 11

OXIDATION-REDUCTION REACTIONS

Chapter Check-In

- ❑ Learning how oxidation numbers are assigned
- ❑ Seeing how oxidation numbers change in oxidation-reduction reactions
- ❑ Using oxidation numbers to balance equations

Oxidation-reduction reactions are some of the most important chemical reactions. **Redox** reactions, as they are called, are the energy-producing reactions in industry as well as in the body. The core of a redox reaction is the passing of one or more electrons from one species to another. The species that loses electrons is said to be oxidized, and the species gaining electrons is reduced. These are old terms, but they are still used today. **Oxidation** and **reduction** occur simultaneously.

Oxidation numbers are assigned to each element in a chemical reaction to help us learn which element is oxidized and which is reduced. If, in a reaction, the oxidation number of an element increases (becomes more positive), the element is being oxidized. On the other hand, if the oxidation number of an element decreases, the element is being reduced. The changes in oxidation numbers are also used to balance redox equations. The goal is to keep the total number of electrons lost in the oxidation equal to the total number gained in the reduction. Clearly, the study of oxidation-reduction reactions should begin by learning about oxidation numbers.

Oxidation Numbers

Ions have an electrical charge—negative if they have gained electrons or positive if they have lost electrons. The existence of ions suggests a transfer of electrons from one atom to another giving rise to the positive and negative charges. A useful extension of this concept is to assign hypothetical charges called oxidation numbers to atoms with polar covalent bonds. The general idea is to assign the shared electrons in each bond to the more electronegative element.

As an example, use the water molecule in a standard Lewis diagram, as shown in Figure 11-1.

Figure 11-1 Lewis structure for H_2O.

Because oxygen is more electronegative than hydrogen, for the purpose of assigning oxidation numbers, it is assumed that all four electrons in the two covalent bonds are associated completely with the oxygen atom. (See Figure 11-2.)

Figure 11-2 Assignment of oxidation numbers.

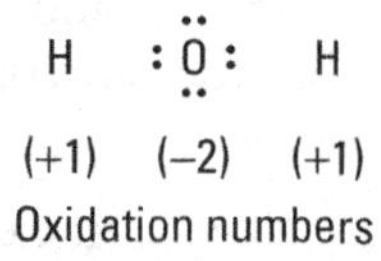

The resulting hypothetical electrical charges are the oxidation numbers, which are shown in parentheses to remind you that they are conceptual rather than real. The atoms in H_2O are not ions.

Four rules apply when assigning oxidation numbers to atoms. First, the oxidation number of each atom in a pure element is defined as zero. Therefore, elemental carbon (graphite or diamond) has an oxidation number of 0, as does an atom in metallic iron, or each of the two hydrogen atoms in the H_2 molecule:

$$\overset{(0)}{C} \quad \overset{(0)}{Fe} \quad \overset{(0)}{H_2}$$

A single-atom ion is assigned an oxidation number equal to its electrical charge. Examples are sodium or iron ions, the latter occurring in two oxidation states:

$$\overset{(+1)}{Na^{+}} \quad \overset{(+2)}{Fe^{2+}} \quad \overset{(+3)}{Fe^{3+}}$$

A multiple-atom molecule or ion must have oxidation numbers that sum to the electrical charge of the group of atoms. A neutral molecule has oxidation numbers adding to zero. Therefore, the oxidation numbers of the 1 nitrogen and 3 hydrogen atoms of the neutral NH_3 ammonia molecule sum to 0:

$$\overset{(-3)}{N} \quad \overset{(+1)}{H_3}$$

whereas the oxidation numbers of the 1 carbon and 3 oxygens of the charged CO_3^{2-} carbonate ion sum to –2:

$$\overset{(+4)}{C} \quad \overset{(-2)}{O_3}$$

For each bond between two different elements, the shared electrons are assigned to the element of greater electronegativity, which was nitrogen in the NH_3 example and oxygen in the CO_3^{2-} example. Oxygen usually has an oxidation number of –2; the halogens are commonly –1; hydrogen is almost always +1; and the alkali metals are +1.

Problem 34: What is the oxidation number of nitrogen in magnesium nitride (Mg_3N_2) and in nitric acid (HNO_3)?

Electron Transfer

The concept of oxidation arises from the combination of elemental oxygen with other elements to form **oxides,** as in this example using aluminum:

$$4Al + 3O_2 \rightarrow 2Al_2O_3$$

With oxidation numbers inserted as superscripts, this reaction is written

$$\overset{(0)}{4Al} + \overset{(0)}{3O_2} \rightarrow 2\overset{(+3)}{Al_2}\overset{(-2)}{O_3}$$

to show that both elements change oxidation numbers. Because the oxidation numbers changed, an **oxidation-reduction reaction** is defined as one in which electrons are transferred between atoms. In the example, each oxygen atom has gained two electrons, and each aluminum has lost three electrons.

In an electron transfer reaction, an element undergoing oxidation loses electrons, whereas an element gaining electrons undergoes reduction. In the aluminum-oxygen example, the aluminum was oxidized, and the oxygen was reduced because *every electron transfer reaction involves simultaneous oxidation and reduction.* These reactions are frequently called redox reactions.

A subtlety deserving your close attention is that the **oxidizing agent** (in the example, oxygen) is reduced, whereas the **reducing agent** (in the example, aluminum) is oxidized.

Because chemists have defined oxidation in terms of electron transfer, it is quite unnecessary for redox reactions to have oxygen as the oxidizing agent. Study the next example of metallic zinc reacting with chlorine gas to form zinc chloride:

$$\underset{Zn}{(0)} \quad + \quad \underset{Cl_2}{(0)} \quad \rightarrow \quad \underset{ZnCl_2}{(+2)(-1)}$$

The oxidizing agent that gains electrons is chlorine, and the reducing agent that loses electrons is zinc.

A valuable generalization is that the nonmetals in the upper right region of the periodic table are strong oxidizing agents. The metals in their elemental state are strong reducing agents, as is hydrogen gas.

Problem 35: In the following redox reaction, identify the element that is oxidized, the element that is reduced, the oxidizing agent, and the reducing agent.

$$\underset{\text{iodine pentoxide}}{I_2O_5} + \underset{\text{carbon monoxide}}{5CO} \rightarrow \underset{\text{iodine}}{I_2} + \underset{\text{carbon dioxide}}{5CO_2}$$

Balancing Equations

A chemical reaction is said to be balanced when the number of atoms of each element is equal in the reactants and products. Because of the conservation of matter, equations are always balanced. You cannot represent the reaction of hydrogen and oxygen as

$$H_2 + O_2 \rightarrow H_2O \text{ (unbalanced)}$$

because there are more oxygen atoms on the left side. The reaction is correctly written

$$2H_2 + O_2 \rightarrow 2H_2O \text{ (balanced)}$$

with exactly four hydrogen atoms and two oxygen atoms on each side. The coefficients of the three substances are chosen so the reaction is properly balanced.

Although the brief reactions described to this point may be quickly balanced by inspection or trial-and-error, chemistry is rich in complicated reactions that cannot be intuitively balanced. An example is the use of concentrated nitric acid to dissolve copper metal. Here is the unbalanced reaction without any coefficients:

$$Cu(s) + H^+(aq) + NO_3^-(aq) \rightarrow Cu^{2+}(aq) + NO_2(g) + H_2O\ (l)$$

Take several minutes and try to balance this equation; the multitude of elements involved makes it a challenging exercise.

Fortunately, such complicated reactions usually involve oxidation and reduction, and the oxidation numbers of each element make it much easier to determine the coefficients for a balanced reaction. First, assign oxidation numbers to the elements in each substance. Examine only the elements that change their oxidation number and insert coefficients so the number of electrons lost equals the number of electrons gained. Then modify any coefficients so the other elements that don't change oxidation number also balance. Finally, check that the electrical charges and the number of atoms of the elements are equal on both sides of the reaction.

As an example, use that method to balance the preceding reaction involving copper and nitric acid. Begin by writing oxidation numbers for all the elements in each substance, as follows:

$$\begin{array}{ccccccccccc} 0 & & 1+ & & 5+2- & & 2+ & & 4+2- & & 1+2- \\ Cu & + & H^+ & + & NO_3^- & \rightarrow & Cu^{2+} & + & NO_2 & + & H_2O \end{array}$$

The oxidation number remains constant for hydrogen and oxygen, and you initially can disregard these elements. Only the copper and nitrogen show changing oxidation numbers:

loses 2 electrons

$$\mathrm{Cu}^{(0)} + \overset{(5+)}{\mathrm{N}}\left(\text{in } \mathrm{NO_3^-}\right) \rightarrow \mathrm{Cu}^{(2+)}\,\overset{(4+)}{\mathrm{N}}\left(\text{in } \mathrm{NO_2}\right)$$

gains 1 electron

The balanced reaction must have equal loss and gain of electrons, so there must be two times as many N atoms as Cu atoms. Insert a coefficient of 2 in front of NO_3^- and NO_2:

$$\mathrm{Cu} + 2\,\mathrm{N}\left(\text{in } \mathrm{NO_3^-}\right) \rightarrow \mathrm{Cu}^{2+} + 2\,\mathrm{N}\left(\text{in } \mathrm{NO_2}\right)$$

The electron gain and loss are now balanced:

$$\mathrm{Cu}(s) + \mathrm{H}^+(aq) + 2\,\mathrm{NO_3^-}(aq) \rightarrow \mathrm{Cu}^{2+}(aq) + 2\,\mathrm{NO_2}(g) + \mathrm{H_2O}(l)$$

Then placing a coefficient of 4 in front of H+ (*aq*) and 2 in front of H_2O (*l*) will give the correctly balanced final equation:

$$\mathrm{Cu}(s) + 4\,\mathrm{H}^+(aq) + 2\,\mathrm{NO_3^-}(aq) \rightarrow \mathrm{Cu}^{2+}(aq) + 2\,\mathrm{NO_2}(g) + 2\,\mathrm{H_2O}(l)$$

The preceding technique for balancing reactions is useful because you begin by considering only the elements involved in electron transfer.

Problem 36: Use oxidation numbers to balance this redox reaction.

$$\mathrm{MnO_4^-}(aq) + \mathrm{H}^+(aq) \rightarrow \mathrm{MnO_2}(s) + \mathrm{O_2}(g) + \mathrm{H_2O}(l)$$

Chapter Check-Out

1. The loss of electrons by a species in a reaction is called ______.

   a. oxidation
   b. redox
   c. reduction

2. What is the oxidation number of copper in CuO?

   a. +1
   b. +2
   c. 0

3. In the following reaction, which element is oxidized, and which element is reduced?

$$Cu(s) + 2\ AgNO_3(aq) \rightarrow Cu(NO_3)_2(aq) + 2\ Ag(s)$$

   a. Cu is reduced; Ag is oxidized.
   b. Ag is reduced; N is oxidized.
   c. Cu is oxidized; Ag is reduced.

4. What is the coefficient placed in front of Cr to correctly balance this equation involving oxidation and reduction?

$$Cr(s) + Sn^{4+}(aq) \rightarrow Cr^{3+}(aq) + Sn^{2+}(aq)$$

   a. 2
   b. 3
   c. 4

5. Which one of the following has an incorrectly assigned oxidation number?

   a. Na_2O, Na is Na^+
   b. H_2O, H is H^+
   c. Fe_2O_3, Fe is Fe^{2+}

Answers: **1.** a **2.** b **3.** c **4.** a **5.** C

Chapter 12
ELECTROCHEMISTRY

Chapter Check-In

- ❑ Learning about the voltaic cell
- ❑ Determining the voltage of a cell
- ❑ Using electricity to cause chemical reactions; the electrolytic cell

Have you ever wondered how a battery works? You can find out how in this chapter. In Chapter 11, you learned how oxidation-reduction reactions transfer electrons from one species to another. Batteries use oxidation-reduction reactions, but they are carefully designed so the flow of electrons takes place through a conducting wire. The first battery was made in 1796 by Alessandro Volta, and batteries are commonly called voltaic cells in his honor. There are many different ways to construct a voltaic cell, but in all cases, two different chemical species must be used. The voltage of the cell depends on which species are used.

If a chemical reaction can make electricity, it should not be surprising to learn that electricity can make a chemical reaction. Using an electric current to cause a chemical reaction is called **electrolysis**, a technique widely used to separate elements from their compounds. For example, pure sodium metal (Na) and chlorine gas (Cl_2) are obtained by passing electricity through molten sodium chloride (NaCl). The study of the interplay of electricity and oxidation-reduction reactions is called electrochemistry.

Electrochemical Cells

You have seen that electron transfer occurs in many chemical reactions. This chapter describes how such redox reactions can generate a flow of electricity and, conversely, how electrical currents can induce chemical reactions.

A device that uses a chemical reaction to produce or use electricity is an **electrochemical cell,** also known as a **voltaic cell.** Because the liquid state allows reactions to occur more readily than in either solids or gases, most electrochemical cells are built using a liquid called an **electrolyte**, a solution that contains ions and conducts electricity. This word has previously been mentioned with regard to ionic dissociation. Pure, distilled water is a very poor conductor of electricity, but a high concentration of dissolved ions leads to high conductivity. That is why acids, bases, and salts that ionize to a high degree are referred to as strong electrolytes, while those that ionize only slightly are referred to as weak electrolytes.

A simple electrochemical cell can be made from two test tubes connected with a third tube (the crossbar of the "H"), as shown in Figure 12-1. The hollow apparatus is filled by simultaneously pouring different solutions into the two test tubes, an aqueous solution of zinc sulfate into the left tube and a copper (II) sulfate solution into the one on the right. Then a strip of zinc metal is dipped into the $ZnSO_4$ solution; a piece of copper is inserted into the $CuSO_4$ solution; and the two ends of the metal strips are connected by wires to a voltmeter. The lateral connecting tube allows ionic migration necessary for a closed electrical circuit. The voltmeter will show the electrical potential of 1.10 volts, which leads to the movement of electrons in the wire from the zinc **electrode** toward the copper electrode.

The electric current is caused by a pair of redox reactions. At the zinc electrode, the metallic zinc is slowly being ionized by an oxidation reaction:

$$\underset{\text{metal}}{Zn(s)} \rightarrow \underset{\text{ion}}{Zn^{2+}(aq)} + \underset{\text{electrons}}{2\,e^-}$$

An electrode at which oxidation occurs is called an **anode;** it strongly attracts negative ions in the solution, and such ions are consequently called **anions.**

Simultaneously, a reduction reaction at the copper **cathode** causes Cu^{2+} **cations** to be deposited onto the electrode as copper metal:

$$\underset{\text{ion}}{Cu^{2+}(aq)} + \underset{\text{electrons}}{2\,e^-} \rightarrow \underset{\text{metal}}{Cu(s)}$$

Because negatively charged electrons are flowing from the anode to the cathode, the anode becomes the positive electrode. The cathode is, therefore, the negative electrode.

Figure 12-1 A voltaic cell.

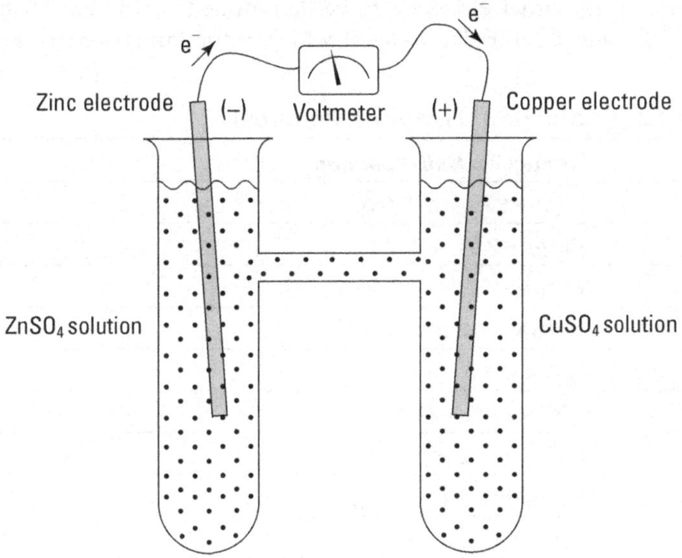

Adding the reactions at the two electrodes gives

$$Zn(s) + Cu^{2+}(aq) \rightarrow Cu(s) + Zn^{2+}(aq)$$

which is the overall redox reaction in the zinc-copper cell.

Electrode Potential

The potential difference, which is measured in *volts* (v), depends upon the particular substances constituting the electrodes. For any electric cell, the total potential is the sum of those produced by the reactions at the two electrodes:

$$EMF_{cell} = EMF_{oxidation} + EMF_{reduction}$$

The **EMF** denotes **electromotive force**, another name for electrical potential.

Chemists have measured the voltages of a great variety of electrodes by connecting each in a cell with a standard hydrogen electrode, which is hydrogen gas at 1 atmosphere bubbling over a platinum wire immersed in 1 M $H^+(aq)$. This standard electrode is arbitrarily assigned a potential of 0 volts, and measurement of the EMF of the complete cell allows the potential of the other electrode to be determined. Table 12-1 lists some standard potentials for electrodes at which reduction is occurring.

Table 12-1 Standard Electrode Potentials

Volts	*Reduction Half-Reaction*
2.87	$F_2(g) + 2\ e^- \rightarrow 2\ F^-(aq)$
1.36	$Cl_2(g) + 2\ e^- \rightarrow 2\ Cl^-(aq)$
1.20	$Pt^{2+}(aq) + 2\ e^- \rightarrow Pt(s)$
0.92	$Hg^{2+}(aq) + 2\ e^- \rightarrow Hg(l)$
0.80	$Ag^+(aq) + e^- \rightarrow Ag(s)$
0.53	$I_2(s) + 2\ e^- \rightarrow 2\ I^-(aq)$
0.34	$Cu^{2+}(aq) + 2\ e^- \rightarrow Cu(s)$
0	$2\ H^+(aq) + 2\ e^- \rightarrow H_2(g)$
-0.13	$Pb^{2+}(aq) + 2\ e^- \rightarrow Pb(s)$
-0.26	$Ni^{2+}(aq) + 2\ e^- \rightarrow Ni(s)$
-0.44	$Fe^{2+}(aq) + 2\ e^- \rightarrow Fe(s)$
-0.76	$Zn^{2+}(aq) + 2\ e^- \rightarrow Zn(s)$
-1.66	$Al^{3+}(aq) + 3\ e^- \rightarrow Al(s)$
-2.71	$Na^+(aq) + e^- \rightarrow Na(s)$
-2.87	$Ca^{2+}(aq) + 2\ e^- \rightarrow Ca(s)$
-2.91	$K^+(aq) + e^- \rightarrow K(s)$
-3.04	$Li^+(aq) + e^- \rightarrow Li(s)$

Near the middle of the list, you will see 0 volts arbitrarily assigned to the standard hydrogen electrode; all other potentials are relative to the hydrogen half-reaction. The voltages are given signs appropriate for a reduction reaction. For oxidation, the sign is reversed; thus, the oxidation half-reaction,

$$Pt(s) \rightarrow Pt^{2+}(aq) + 2\ e^-$$

has an EMF of −1.20 volts, the opposite given in Table 12-1. Look this up in the chart to be sure that you understand.

Consider how these standard potentials are used to determine the voltage of an electric cell. In the zinc-copper cell described earlier, the two half-reactions must be added to determine the cell EMF. (See Table 12-2.)

Table 12-2 Zinc-Copper Cell

Half-Reaction	***Type***	***Electrode***	***Potential***
$Zn(s) \rightarrow Zn^{2+}(aq) + 2\ e^-$	Oxidation	Anode	0.76 volt
$Cu^{2+}(aq) + 2\ e^- \rightarrow Cu(s)$	Reduction	Cathode	0.34 volt

The complete zinc-copper cell has a total potential of 1.10 volts (the sum of 0.76v and 0.34v). Notice that the sign of the potential of the zinc anode is the reverse of the sign given in the chart of standard electrode potentials (see Table 12-1) because the reaction at the anode is oxidation.

In the chart of standard electrode potentials (see Table 12-1), reactions are arranged in order of their tendency to occur. Reactions with a positive EMF occur more readily than those with a negative EMF. The zinc-copper cell has an overall EMF of +1.10 volts, so the dissolution of zinc and deposition of copper can proceed.

Calculate the total potential of a similar cell with zinc and aluminum electrodes. Table 12-3 shows the two pertinent half-reactions.

Table 12-3 Aluminum-Zinc Cell

Half-Reaction	***Type***	***Electrode***	***Potential***
$2\ Al(s) \rightarrow 2\ Al^{3+}(aq) + 6\ e^-$	Oxidation	Anode	1.66 volts
$3\ Zn^{2+}(aq) + 6\ e^- \rightarrow 3\ Zn(s)$	Reduction	Cathode	-0.76 volt

Such a cell with zinc and aluminum electrodes would have an overall potential of +0.90 volt, with aluminum being dissolved and zinc metal being deposited out of solution.

If you select any two half-reactions from the chart of standard electrode potentials, the half-reaction higher on the list will proceed as a reduction, and the one lower on the list will proceed in the reverse direction, as an oxidation. ***Beware:*** Some references give standard electrode potentials for oxidation half-reactions, so you have to switch "higher" and "lower" in the rule stated in the preceding sentence, though this is not common.

Problem 37: Some silver mines dump shredded iron cans into ponds containing dissolved silver salts. Write the two redox half-reactions and the overall balanced reaction that explains the deposition of silver from the solution.

Problem 38: Considering only the elements in the chart of standard electrode potentials (see Table 12-1), which pair can make a battery with the greatest voltage? What would the voltage be?

Faraday's Laws

The electrochemical cell with zinc and copper electrodes had an overall potential difference that was positive (+1.10 volts), so the spontaneous chemical reactions produced an electric current. Such a cell is called a *voltaic cell.* In contrast, *electrolytic cells* use an externally generated electrical current to produce a chemical reaction that would not otherwise take place.

An instance of such electrolysis is the decomposition of water to elemental hydrogen and oxygen. The pertinent half-reactions are given in Table 12-4.

Table 12-4 Electrolysis of Water

Half-Reaction	*Type*	*Electrode*	*Potential*
$4\ H_2O(l) + 4\ e^- \rightarrow 2\ H_2(g) + 4\ OH^-(aq)$	Reduction	Cathode	-0.83 volt
$2\ H_2O(l) \rightarrow O_2(g) + 4\ H^+(aq) + 4\ e^-$	Oxidation	Anode	-0.40 volt

The overall reaction for the electrolysis of water is given by adding the two half-reactions to obtain

$$2\ H_2O(l) \rightarrow 2\ H_2(g) + O_2(g)$$

with an overall potential of −1.23 volts. With a negative potential, it requires an externally imposed electrical current to decompose water by the reaction shown. Figure 12-2 shows two platinum electrodes in water containing a little salt or acid so that the solution can conduct electricity.

Figure 12-2 The electrolysis of water.

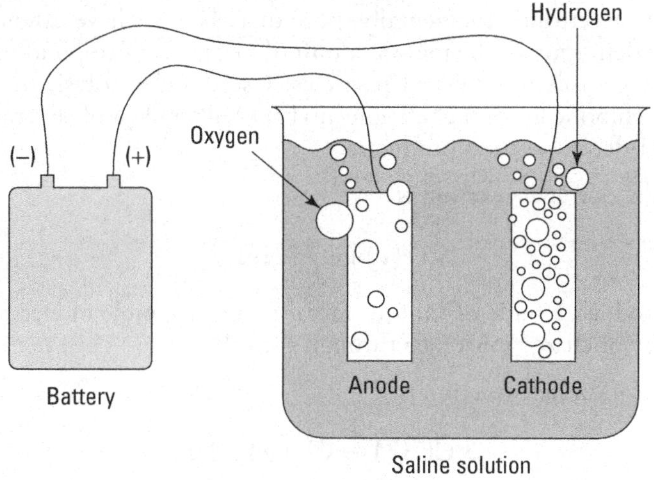

The reduction at the cathode yields H_2 gas, and the oxidation at the anode yields O_2 gas. Notice that the figure shows that the volume of hydrogen is twice the volume of oxygen—look at the bubbles. The molar coefficients in the decomposition reaction imply 2 volumes of H_2 gas for each 1 volume of O_2 gas.

Electrolysis is used to decompose many compounds into their constituent elements. You have seen this process with water. Another example is the electrolysis of molten sodium chloride to yield molten sodium metal and chlorine gas:

$$2\ \mathrm{NaCl}(l) \rightarrow 2\ \mathrm{Na}(l) + \mathrm{Cl_2}(g)$$

Chemists throughout the nineteenth century discovered new elements from the electrolytic decomposition of many compounds.

The quantitative laws of electrochemistry were discovered by Michael Faraday of England. His 1834 paper on electrolysis introduced many of the terms that you have seen throughout this book, including *ion, cation, anion, electrode, cathode, anode,* and *electrolyte.* He found that the mass of a substance produced by a redox reaction at an electrode is proportional to the quantity of electrical charge that has passed through the electrochemical cell. For elements with different oxidation numbers, the same quantity of electricity produces fewer moles of the element with the higher oxidation number.

The basic unit of electrical charge used by chemists is appropriately called a **faraday**, which is defined as the charge on one mole of electrons (6×10^{23} electrons). Incidentally, note that chemists have extended the original definition of the mole as a unit of mass to a corresponding number (Avogadro's number) of particles. Use the electrolysis of molten sodium chloride to see the relationship between faradays of electricity and moles of decomposition products.

The reduction half-reaction is

$$Na^+(l) + e^- \rightarrow Na(l)$$

so to produce 1 mole of sodium metal requires 1 mole of electrons, so 1 faraday of charge must pass through the cell.

The oxidation half-reaction is

$$2\ Cl^-(l) \rightarrow Cl_2(g) + 2\ e^-$$

and to produce 1 mole of chlorine gas, 2 faradays of electric charge must pass through the apparatus. Notice how the number of electrons in redox reactions determines the quantity of electricity needed for the reaction.

These half-reactions sum to the overall reaction in the electrolytic cell:

$$2\ Na^+(l) + 2\ Cl^-(l) \rightarrow 2\ Na(l) + Cl_2(g)$$

The passage of 2 faradays of charge yields 2 moles of sodium metal and 1 mole of chlorine gas.

The first of **Faraday's laws** states that the mass of substance produced is proportional to the quantity of electricity. To apply this law to the NaCl example, where 1 mole of Cl_2 was produced by 2 faradays, means that to produce 10 moles of Cl_2 requires the passage of 20 faradays through the apparatus.

The second of Faraday's laws states that a given quantity of electricity produces fewer moles of substances with higher oxidation numbers. Compare the reduction of sodium and calcium ions:

$$Na^+ + e^- \rightarrow Na\ (\text{1 mole of electrons})$$

$$Ca^{2+} + 2\ e^- \rightarrow Ca\ (\text{2 moles of electrons})$$

It requires twice as much electricity to produce 1 mole of calcium metal as 1 mole of sodium metal.

The electrolytic decomposition of sodium chloride and calcium oxide appear similar:

$$2\ NaCl(l) \rightarrow 2\ Na(l) + Cl_2(g)$$
$$2\ CaO(l) \rightarrow 2\ Ca(l) + O_2(g)$$

but the NaCl decomposition requires the transfer of only half as many electrons as does the CaO decomposition. For the NaCl electrolysis, it was previously calculated that the passage of 20 faradays of electric charge produced 20 moles of sodium metal and 10 moles of chlorine gas. The same amount of electric charge passing through the CaO cell yields only 10 moles of calcium metal and 5 moles of oxygen gas.

The reduction half-reaction with balanced coefficients:

$$2\ Ca^{2+}(l) + 2\ e^{-} \rightarrow Ca(l)$$

The oxidation half-reaction with balanced coefficients:

$$2\ O^{2-}(l) \rightarrow O_2(g) + 2\ e^{-}$$

Problem 39: Aluminum metal is produced by the electrolysis of molten cryolite, Na_3AlF_6. How many faradays of electric charge are needed to produce 1 kilogram of aluminum?

Chapter Check-Out

1. Oxidation occurs at the ______.

a. anode
b. cathode
c. electrolyte

2. Consulting the standard electrode potentials (see Table 12-1), what is the voltage for a cell using this reaction?

$$Zn(s) + Ni^{2+}(aq) \rightarrow Zn^{2+}(aq) + Ni(s)$$

a. +0.50 v
b. +1.02 v
c. –1.02 v

3. Consulting the standard electrode potentials (see Table 12-1), what is the voltage for a cell using this reaction?

$$Cu(s) + 2\ Ag^{+}(aq) \rightarrow Cu^{2+}(aq) + 2\ Ag(s)$$

a. +1.14 v
b. –0.46 v
c. +0.46 v

4. How many faradays of electricity are needed to reduce one mole of Cr^{3+} to Cr?

$$Cr^{3+}(aq) + 3\ e^{-} \rightarrow Cr(s)$$

a. 1
b. 2
c. 3

5. In the electrolysis of molten NaCl, if 23.0 grams of sodium metal is obtained when 1.0 faraday of electricity passes through the cell, how many grams of sodium are obtained if 1.5 faradays pass through the cell?

a. 23.0 g of Na
b. 34.5 g of Na
c. 15.3 g of Na

Answers: **1.** a **2.** a **3.** c **4.** c **5.** b

Chapter 13
EQUILIBRIUM

Chapter Check-In

- ❑ Understanding equilibrium in chemical systems
- ❑ Predicting how equilibrium responds to changes in temperature, concentration, and pressure
- ❑ Learning how to express an equilibrium state quantitatively

Sometimes, when a chemical reaction takes place, it proceeds for a period of time and then seems to stop before all the reactants are consumed. But the reaction does not actually stop. Instead, the reaction reaches a point of chemical equilibrium in which the reverse reaction is converting products into reactants as fast as products are formed in the forward reaction. At equilibrium, with both the forward and reverse reactions taking place at the same rate, the concentration of every species no longer changes.

Every reaction has a point in which equilibrium is established. For many reactions, it occurs at the point when essentially all reactants are converted to products; for practical purposes, scientists say that the reaction has gone to completion. But for other reactions, equilibrium occurs when only part of the reactants are converted into products. In these cases and with enough information, it is possible to calculate the concentration of one or more species at equilibrium. In addition, it is possible to predict how the equilibrium will be affected by a change in temperature or an increase or decrease in concentration of a reactant or product. The state of equilibrium is especially important in solutions, and there are many vital equilibria in the chemistry of the body.

Two Reaction Directions

In Chapter 10, your attention was drawn to the idea of equilibrium (a stable, balanced condition resulting from two opposing reactions) and the central role that it plays in chemistry.

You are aware that the melting of ice can be represented by the equation

$$H_2O(s) \rightarrow H_2O(l)$$

and that the freezing of water can be represented by the reverse equation

$$H_2O(l) \rightarrow H_2O(s)$$

Either of these unilateral reactions is written with the implication that the reactant on the left is completely converted to the product on the right. The situation in which both states of H_2O are in equilibrium is shown by the reversible reaction

$$H_2O(l) \rightleftarrows H_2O(l)$$

where the two arrows mean that some H_2O molecules are participating in the forward (melting) reaction and other molecules are simultaneously participating in the reverse (freezing) reaction. Therefore, equilibrium is the stable situation resulting from two offsetting reactions. At a pressure of 1 atmosphere and a temperature of 0°C, both solid ice and liquid water are stable and will coexist. Notice that this equilibrium condition can be reached from either side; it can begin with either pure ice at –10°C or pure water at 20°C.

Whether you warm such ice or cool such water, the second phase will appear at 0°C.

The second example demonstrates another aspect of equilibrium by using the transformations between dinitrogen tetroxide and nitrogen dioxide.

$$\underset{\text{dinitrogen textroxide}}{N_2O_4(g)} \rightleftarrows \underset{\text{nitrogen dioxide}}{2\,NO_2(g)}$$

N_2O_4 is a colorless gas, whereas NO_2 is a dark reddish-brown gas. Their relative abundances are a function of temperature because N_2O_4 dominates at room temperature, and NO_2 dominates at higher temperatures. At any one temperature, both gases are present in a mixture, and the color of the mixture allows an estimation of the ratio of the two nitrogen oxides. If a glass vessel containing them is colorless or pale, N_2O_4 exceeds NO_2. Warming that container would cause the color to slowly darken as N_2O_4 is converted to NO_2 until the ratio of the two species is appropriate for the higher temperature. (See Figure 13-1.)

Figure 13-1 Temperature and the N_2O_4-NO_2 mixture.

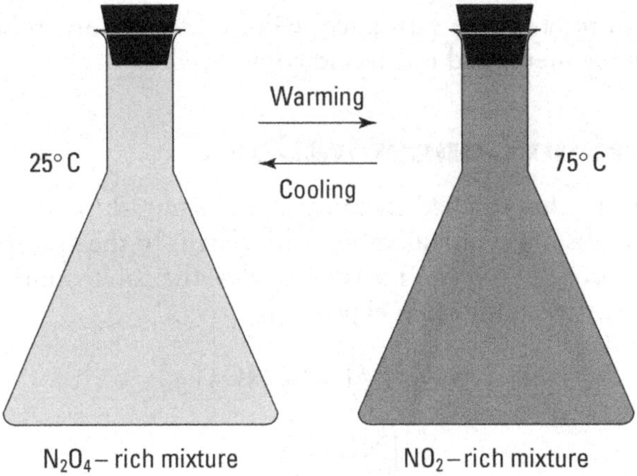

Then the color ceases to change and remains at the new, darker hue. The color change in this gaseous reaction allows you to readily imagine the pair of reactions involved. The experimental fact that warming the gas mixture causes the color to darken shows that temporarily the reaction

$$N_2O_4(g) \rightarrow 2\,NO_2(g)$$

dominates over the reverse reaction. The additional consequence that the color stops getting darker shows that a new chemical equilibrium has been reached:

$$N_2O_4(g) \rightleftarrows 2\,NO_2(g)$$

in which the forward and reverse reactions occur at the same rate. The rate of creation of NO_2 is precisely equal to the rate of its consumption forming N_2O_4.

The next example demonstrates that equilibrium may involve more than two substances.

$$As_4O_6(s) + 6\,C(s) \rightleftarrows As_4(g) + 6\,CO(g)$$

At equilibrium, all four substances will be present: one solid compound, one solid element, one gaseous element, and one gaseous compound.

The final example is the preparation of liquid phosphorus trichloride by the following reaction:

$$P_4(s) + 6\ Cl_2(g) \rightleftarrows 4\ PCl_3(l)$$

At equilibrium, all three substances will be present: one solid element, one gaseous element, and one liquid compound.

Equilibrium Concentrations

In a system that has reached chemical equilibrium, the concentrations of the various substances are quantitatively related. In the example of transformations between the two nitrogen oxides, the concentrations at equilibrium obey this equilibrium expression:

$$N_4O_4(g) \rightleftarrows 2\ NO_2(g)$$

$$\frac{[NO_2]^2}{[N_2O_4]} = K$$

where the value K is the **equilibrium constant.**

For any equilibrium reaction, the ratio of concentrations of the substances on the right to the concentrations of those on the left equals a constant appropriate for that specific reaction. Notice that the ratio is always written with the products over the reactants. Each concentration must be raised to the power of its stoichiometric coefficient in the reaction. Because NO_2 has a coefficient of 2 in the reaction, its concentration must be squared in the preceding expression.

For a generalized reaction written

$$w\,A + x\,B \rightarrow y\,C + z\,D$$

where the lowercase letters represent numerical coefficients for the balanced reaction, the equilibrium constant is calculated

$$\frac{[C]^y[D]^z}{[A]^w[B]^x} = K$$

where the brackets denote concentrations of the various substances.

Therefore, the reaction of nitrogen and hydrogen to yield ammonia

$$N_2(g) + 3\,H_2(g) \rightleftarrows 2\,NH_3(g)$$

obeys this equilibrium equation:

$$\frac{[NH_3]^2}{[N_2][H_2]^3} = K$$

The dissociation of water vapor to hydrogen and oxygen

$$2\,H_2O(g) \rightleftarrows 2\,H_2(g) + O_2(g)$$

follows this equation:

$$\frac{[H_2]^2[O_2]}{[H_2O]^2} = K$$

Of course, the equilibrium constant *K* in the latter equation does not have the same value as the *K* in the equilibrium equation for ammonia. The numerical value of *K* depends on the particular reaction, the temperature, and the units used to describe concentration. For liquid solutions, the concentrations are usually expressed as molarity. For a mixture of gases, the concentration of each molecular species is commonly given either as molarity or as pressure in atmospheres.

Use the ammonia equilibrium,

$$2\,N_2(g) + 3\,H_2(g) \rightleftarrows 2\,NH_3(g)$$

where the concentrations are measured as pressures. At 25°C, one study reported the equilibrium pressures of the three gases. See Table 13-1.

Table 13-1 Ammonia Equilibrium

Gas	*Pressure (atm)*
NH_3	1.74
H_2	0.06
N_2	0.02

From the experimental data in Table 13-1, you can calculate the equilibrium constant for the reaction at 25°C.

$$K = \frac{\left(P_{NH_3}\right)^2}{\left(P_{N_2}\right)\left(P_{H_2}\right)^3} = \frac{(1.74)^2}{(0.02)(0.06)^3} = 7\times10^5$$

Equilibrium constants have been determined for many reactions over a wide range of temperatures. One obvious use is to calculate the concentrations of the various substances at equilibrium. In the ammonia example, if you had been given the equilibrium constant K and the pressures of ammonia and hydrogen, you could have calculated the pressure of nitrogen. (You may want to attempt this simple calculation.)

Another important use of equilibrium constants is to predict the initial direction of a reaction. Most commonly, the original concentrations of substances are unstable with respect to equilibrium; this is necessarily the case if any substance is absent. The reaction will proceed in one direction until equilibrium is attained. Referring to the ammonia example, if initially

$$\frac{\left(P_{NH_3}\right)^2}{\left(P_{N_2}\right)\left(P_{H_2}\right)^3} > 7\times10^5$$

ammonia decomposes to nitrogen and hydrogen with decreasing P_{NH_3} and increasing P_{N_3} and P_{H_2} until the equilibrium pressures are reached. In the converse case, where initially

$$\frac{\left(P_{NH_3}\right)^2}{\left(P_{N_2}\right)\left(P_{H_2}\right)^3} < 7\times10^5$$

nitrogen and hydrogen will combine to form ammonia with decreasing P_{N_2} and P_{H_2} and increasing P_{NH_3} until the values satisfy the equilibrium constant. Notice that only concentrations at equilibrium are stable and unchanging at a given temperature.

Although equilibrium calculations involve the concentrations of dissolved substances and gases, the values for pure liquids and solids are virtually constant and so are usually not incorporated into the equilibrium constant. As an illustration of this point, the reaction

$$CO(g) + H_2O(l) \rightleftarrows CO_2(g) + H_2(g)$$

contains a liquid (H_2O) that has a fixed composition. The equilibrium may be expressed by either of the two equations:

$$K_c = \frac{[CO_2][H_2]}{[CO]} \quad \text{or} \quad K_p = \frac{(P_{CO_2})(P_{H_2})}{(P_{CO})}$$

where the concentration of H_2O does not appear because, as a constant, it is included in the value of the equilibrium constant. As was mentioned earlier, there are different equilibrium constants depending on whether the concentrations are in molarity (K_c) or pressure (K_p).

As one more example of the nonappearance of pure phases in the equilibrium equation, examine the reaction of lithium bromide with hydrogen:

$$LiBr(s) + H_2(g) \rightleftarrows LiH(s) + HBr(g)$$

The equilibrium equation is

$$K = \frac{[HBr]}{[H_2]} = \frac{P_{HBr}}{P_{H_2}}$$

Constant concentrations of the two solids (lithium bromide and lithium hydride) are not included in the value of *K*. Although pure liquids and solids do not appear in the equilibrium expression as commonly written, they must be present as real substances in the actual reaction for the two opposing reactions to be at equilibrium.

The earlier calculations for both acid dissociations and solubility products are special applications of finding concentrations from equilibrium constants.

Problem 40: At 100°C, the halogen reaction

$$Br_2(g) + Cl_2(g) \rightleftarrows 2\ BrCl(g)$$

has an equilibrium constant of 0.15 when the concentrations are expressed in either molarity or atmospheres. A 1.0 liter reaction vessel is filled with 100 grams of Br_2 and 50 grams each of Cl_2 and BrCl. Do an equilibrium calculation to predict the direction of the initial reaction.

Problem 41: At 700°C, the reaction

$$2\ SO_2(g) + O_2(g) \rightleftarrows 2\ SO_3(g)$$

has an equilibrium constant K_p = 3.76 when the concentrations of the gases are expressed in atmospheres. If $P_{SO_2} = 0.2$ atm and $P_{O_2} = 0.5$ atm, what is the pressure of SO_3? Assume that the gases are at equilibrium.

Le Chatelier's Principle

A valuable guide is available to assist you in estimating how chemical equilibrium will shift in response to changes in the conditions of the reaction, such as a modification of temperature or pressure. The French chemist Henri Le Chatelier realized in 1884 that if a chemical system at equilibrium is disturbed, the system would adjust itself to minimize the effect of the disturbance. This qualitative reasoning tool is cited as **Le Chatelier's principle.**

Begin by considering how an equilibrium system adjusts to a change in the concentration of any substance. At equilibrium, the concentrations of all substances are fixed, and their ratio yields the equilibrium constant. Le Chatelier's principle tells you that changing the concentration of a substance causes the system to adjust to minimize the change in that substance. The decomposition of carbonyl bromide provides an illustration:

$$COBr_2(g) \rightleftarrows CO(g) + Br_2(g)$$

If the three gases in the reaction were at equilibrium and you then increased the carbon monoxide concentration, some Br_2 would combine with added CO to produce $COBr_2$ and thereby minimize the increase in CO. Alternatively, if you decrease the CO concentration, some $COBr_2$ would decompose to produce CO and Br_2 and thereby minimize any decrease in CO. Notice how the concentrations of all constituents shift to counteract the imposed change in a single substance. Of course, this shift does not affect the value of the equilibrium constant. Only a change in temperature can do that.

Notice how a change in pressure would affect the equilibrium reaction.

$$\underbrace{N_2(g)+3\,H_2(g)}_{\text{4 volumes}} \rightleftarrows \underbrace{2\,NH_3(g)}_{\text{2 volumes}}$$

which has an equilibrium constant at standard temperature and pressure that is calculated as

$$\frac{\left(P_{NH_3}\right)^2}{\left(P_{N_2}\right)\left(P_{H_2}\right)^3}=K$$

An increase in pressure will, according to Le Chatelier, cause the equilibrium to shift to minimize the pressure increase. Because the equilibrium reaction has more relative volumes on the left side, the pressure increase would be minimized by some N_2 and H_2 (total of four volumes) combining to form NH_3 (two volumes). Although the relative pressures of the gases have changed, the equilibrium constant still equals *K*. Conversely, a decrease in pressure would be minimized by the dissociation of some NH_3 (two volumes) to form N_2 and H_2 (four volumes).

The only reactions that are significantly affected by pressure are reactions involving gases in which the stoichiometric coefficients of the gases add to different values on the two sides of the reaction. Pressure, therefore, would not affect the equilibrium of

$$N_2(g)+O_2(g)\rightleftarrows 2\,NO(g)$$

which has two volumes on each side. But pressure would affect the equilibrium for

$$2\,CO(g)+O_2(g)\rightleftarrows 2\,CO_2(g)$$

which has three volumes on the left and only two on the right. In this latter example, an increase of pressure induces the forward reaction, and a decrease in pressure causes the reverse reaction. Notice that the effect of varying pressure is to cause the concentrations of the various gases to shift, without any change in the equilibrium constant.

A change in temperature, however, does force a change in the equilibrium constant. Most chemical reactions exchange heat with the surroundings.

A reaction that gives off heat is classified as **exothermic**, whereas a reaction that requires the input of heat is said to be **endothermic**. (See Table 13-2.) A simple example of an endothermic reaction is the vaporization of water:

$$H_2O(l) \rightarrow H_2O(g)$$

which absorbs 40.7 kilojoules per mole. The converse condensation reaction

$$H_2O(g) \rightarrow H_2O(l)$$

is exothermic because it releases 40.7 kilojoules per mole. A change of state is not required for heat to be involved in a reaction. The combustion of methane

$$\text{Heat} + COBr_2(s) \rightarrow CO(g) + Br_2(g)$$

involves only gases, yet this endothermic reaction absorbs heat.

Table 13-2 Thermal Classes of Reactions

Role of Heat	***Description***
Reactants + Heat → Products	Endothermic
Reactants → Products + Heat	Exothermic

In a system at chemical equilibrium, there are always two opposing reactions, one endothermic and the other exothermic.

You can now consider how a change in temperature affects chemical equilibrium. In accordance with Le Chatelier's principle, the equilibrium constant changes to minimize the change in temperature. For endothermic reactions, an increase in temperature can be minimized by utilizing some of the heat to convert reactants to products, shifting the equilibrium to the right side of the reaction and increasing the value of *K*. For exothermic reactions, an increase in temperature can be minimized by using some of the heat to convert "products" into "reactants" and shifting the equilibrium toward the left side, reducing the value of *K*.

For a chemical system at equilibrium, an increase in temperature favors the endothermic reaction, whereas a decrease in temperature favors the exothermic reaction. The equilibrium reaction written as

$$\text{Heat} + \text{COBr}_2(g) \rightleftarrows \text{CO}(g) + \text{Br}_2(g)$$

is endothermic when proceeding to the right and exothermic when proceeding to the left. Its equilibrium constant, given by

$$K = \frac{[\text{CO}][\text{Br}_2]}{[\text{COBr}_2]}$$

must increase if the temperature increases. Conversely, a lowering of the temperature will cause *K* to decrease.

Please realize that the effect of temperature on the equilibrium constant depends on which of the two opposing reactions is exothermic and on which is endothermic. You must have information on the heat of a reaction before you can apply Le Chatelier's principle to judge how temperature alters the equilibrium.

The next two practice problems refer to the following reaction, which is endothermic in the forward direction.

$$\text{N}_2\text{O}_4(g) \rightleftarrows 2\,\text{NO}_2(g)$$

Problem 42: How would an increase in total confining pressure affect the masses of the two nitrogen oxides?

Problem 43: How would an increase in temperature affect the masses of the two nitrogen oxides?

Chapter Check-Out

1. Which of the following is not consistent with the state of equilibrium?
   - **a.** The forward and reverse reactions are both occurring.
   - **b.** The forward and reverse reactions are both endothermic.
   - **c.** If an equilibrium is disturbed, it will change and form a new equilibrium.
2. Which substance(s) in the following equilibrium would not appear in the equilibrium equation, which has the form K = products/reactants?

   $$Fe_3O_4(s) + 4\,H_2(g) \rightleftarrows 3\,Fe(s) + 4\,H_2O(g)$$

   - **a.** $Fe_3O_4(s)$ and $H_2(g)$
   - **b.** $Fe_3O_4(s)$ and $Fe(s)$
   - **c.** $H_2(g)$ and $H_2O(g)$
3. Consider the mixture of PCl_5 and $PCl_3 + Cl_2$, which is in equilibrium at 150°C:

   $$PCl_5(g) \rightleftarrows PCl_3(g) + Cl_2(g)$$

   What will happen to the molarity of Cl_2 if the concentration of PCl_5 is suddenly increased?

   - **a.** It will increase.
   - **b.** It will not change.
   - **c.** It will decrease.
4. What is the only thing that can change the value of an equilibrium constant for a given reaction?
   - **a.** a change in the concentration of a reactant
   - **b.** a change in the temperature
   - **c.** a change in the pressure of one or more gaseous reactants
5. A chemical reaction that produces heat as it takes place is ______.
   - **a.** a fast reaction
   - **b.** an endothermic reaction
   - **c.** an exothermic reaction

Answers: **1.** b **2.** b **3.** a **4.** b **5.** c

Chapter 14
THERMODYNAMICS

Chapter Check-In

- ❑ Learning how to determine the amount of heat produced or absorbed as a reaction takes place
- ❑ Learning about entropy and the importance of increasing disorganization in nature
- ❑ Learning to predict whether or not a reaction will take place

Thermodynamics is the study of the energy, principally heat energy, that accompanies chemical or physical changes. Some chemical reactions release heat energy; they are called exothermic reactions, and they have a negative enthalpy change. Others absorb heat energy and are called endothermic reactions, and they have a positive enthalpy change. But thermodynamics is concerned with more than just heat energy. The change in level of organization or disorganization of reactants and products as changes take place is described by the entropy change of the process. For example, the conversion of one gram of liquid water to gaseous water is in the direction of increasing disorder, the molecules being much more disorganized as a gas than as a liquid. The increase in disorder is described as an increase in entropy, and the change in entropy is positive.

Whether a chemical reaction or physical change will occur depends on both the **enthalpy** and **entropy** of the process, which are quantities that can be calculated from tabulated data. Both terms are combined in the **free energy**—the third and most important thermodynamic term. If the change in free energy is negative, the reaction will proceed to the right; this reaction is called a spontaneous reaction. If the sign is positive, the reaction will not proceed as written; this reaction is nonspontaneous. A powerful prediction as to whether a reaction will or will not take place can be made using tabulated data to calculate the change in free energy.

Thermodynamics is a powerful tool for chemists. After all, nature left to its own devices is always moving toward a minimum of potential energy. Thermodynamics tells the chemist in which direction this minimum lies.

Enthalpy

The experimental discovery that almost all chemical reactions either absorb or release heat led to the idea that all substances contain heat. Consequently, the *heat of a reaction* is the difference in the heat contents of the products and reactants:

$$\Delta H = H_{\text{products}} - H_{\text{reactants}}$$

Throughout this book, the Greek letter delta Δ will be used to symbolize change. Chemists use the term enthalpy for the heat content of a substance or the heat of a reaction, so the *H* in the previous equation means enthalpy. The equation states that the change in enthalpy during a reaction equals the enthalpy of the products minus the enthalpy of the reactants. You can consider enthalpy to be chemical energy that is commonly manifested as heat.

Use the decomposition of ammonium nitrate as an example of an enthalpy calculation. The reaction is

$$NH_4NO_3(s) \rightarrow N_2O(g) + 2\,H_2O(g)$$

and the enthalpies of the three compounds are given in Table 14-1.

Table 14-1 Enthalpies of Compounds

Compound	***(kJ/mole)***
NH_4NO_3 (*s*)	-366
N_2O (*g*)	82
H_2O (*g*)	-242

Notice that the enthalpies can be either positive or negative. In general, compounds that release heat when they are formed from their elements have a negative enthalpy, and substances that require heat for their formation have a positive enthalpy. The enthalpy of the decomposition reaction can be calculated as follows:

$$\Delta H = \underbrace{\left[\overset{N_2O}{82} + 2(\overset{H_2O}{-242})\right]}_{\text{products}} - \underbrace{\left[\overset{NH_4NO_3}{-366}\right]}_{\text{reactant}} = -36 \text{ kJ}$$

Observe the doubling of the enthalpy of H_2O (–36 kJ/mole) because this compound has a stoichiometric coefficient of 2 in the reaction. The overall enthalpy of the reaction is –36 kilojoules, which means that the decomposition of 1 mole of ammonium nitrate releases 36 kJ of heat. The release of heat means that this is an exothermic reaction. The sign of the enthalpy of the reaction reveals the direction of heat flow. See Table 14-2.

Table 14-2 Enthalpy and Heat Flow

Sign of ΔH	***Type of Reaction***	***Heat***
Negative	Exothermic	Released
Positive	Endothermic	Absorbed

If you reverse the previous reaction,

$$N_2O(g) + 2\ H_2O(g) \rightarrow NH_4NO_3(s)$$

the sign of the enthalpy of the reaction is reversed:

$$\Delta H = +36 \text{ kJ}$$

The reversed reaction is, therefore, endothermic. It would require the addition of 36 kcal of energy in order to cause the nitrous oxide and water vapor to react to form 1 mole of ammonium nitrate.

The calculations on the ammonium nitrate reaction demonstrate the immense value of tables that list the enthalpies for various substances. The values at 25°C and 1 atm are called *standard enthalpies*. For elements, the standard enthalpy is defined as 0. For compounds, the values are called standard enthalpies of formation because the compounds are considered to be formed from elements in their standard state.

Table 14-3 gives a few values that will be used in subsequent examples and problems. The symbol for standard enthalpies of formation is ΔH_f°, where the superscript denotes standard and the subscript denotes formation. Look up both elemental sulfur and nitrogen to see that the standard enthalpies for elements are 0. Then find the pairs of values for H_2O and CCl_4 (carbon tetrachloride) to learn that the enthalpy depends on the state of matter.

Table 14-3 Standard Enthalpies of Formation (kJ/mole)

Compound	ΔH_f°	*Compound*	ΔH_f°
Al (*s*)	0	$MgCl_2$ (*s*)	-642
$AlCl_3$ (*s*)	-704	MgO (*s*)	-602
Al_2O_3 (*s*)	-1676	$Mg(OH)_2$ (*s*)	-925
CCl_4 (*g*)	-103	N_2 (*g*)	0
CCl_4 (*l*)	-140	NH_3 (*g*)	-46
CO (*g*)	-111	NO (*g*)	90
CO_2 (*g*)	-394	N_2O (*g*)	82
CaF_2 (*s*)	-1220	O_2 (*g*)	0
CaO (*s*)	-635	O_3 (*g*)	143
$Ca(OH)_2$ (*s*)	-987	S (*s*)	0
HCl (*g*)	-92	SO_2 (*g*)	-297
H_2O (*g*)	-242	SO_3 (*g*)	-396
H_2O (*l*)	-286	$ZnCl_2$ (*s*)	-415
H_2S (*g*)	-21	ZnO (*s*)	-348

Use the data in Table 14-3 to calculate the enthalpy change of the following reaction:

$$2\ H_2S(g) + 3\ O_2(g) \rightarrow 2\ H_2O(l) + 2\ SO_2(g)$$

The difference in the heats of formation of the products is given by:

$$\Delta H = [\underset{H_2O}{2(-286)} + \underset{SO_2}{2(-297)}]_{\text{products}} - [\underset{H_2S}{2(-21)} + \underset{O_2}{3(0)}]_{\text{reactants}} = -1124\ \text{kJ}$$

The enthalpy of the reaction is –1124 kilojoules, meaning that the oxidation of 2 moles of hydrogen sulfide yields or releases 1124 kJ of heat. This reaction is exothermic.

Use Table 14-3 for the next two problems.

Problem 44: Calculate the enthalpy change for the following reaction and classify it as exothermic or endothermic.

$$MgCl_2(s) + H_2O(l) \rightarrow MgO(s) + 2\ HCl(g)$$

Problem 45: Calculate the quantity of heat released when 100 grams of calcium oxide react with liquid water to form $Ca(OH)_2$ (*s*).

Energy and Entropy

Many powerful calculations in thermodynamics are based on a few fundamental principles, which are called the laws of **thermodynamics.** Begin by reviewing the two main laws of the field.

The first law of thermodynamics asserts that energy is conserved during any process. The three major forms of energy for chemical purposes are the internal energy of each substance, the external work due to changes in pressure or volume, and the exchange of heat with the surroundings.

The internal energy is sometimes called *chemical energy* because it is the consequence of all the motions of particles and forces between particles: molecules, atoms, nucleons, and electrons.

If no heat flows in or out of a sample of matter, any external work done on or by the sample is precisely offset by the opposite change in internal energy. Expansion against a confining pressure reduces the internal energy, whereas external compression of the system increases the internal energy.

The first law of thermodynamics also tells you that if no work is done on or by the sample—that is, pressure and volume are held constant—any heat flow is counterbalanced by a change in internal energy. An exothermic reaction releasing heat to the surroundings, therefore, is accompanied by a decrease in internal energy, whereas an endothermic reaction has a concomitant increase in internal energy.

The second law of thermodynamics involves entropy, which for our purposes is a statistical measure of the degree of disorder in a chemical system. As an illustration, compare the arrangements of Na^+ and Cl^- ions in both solid and liquid sodium chloride.

Solid sodium chloride has a crystalline structure in which the cations and anions alternate in a repeating pattern. (See Figure 14-1.)

Figure 14-1 The crystal structure of NaCl.

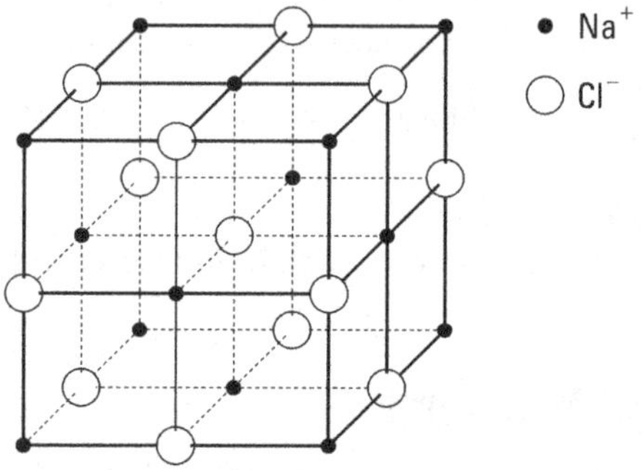

If you examine the NaCl crystal structure closely, you will see that each Cl^- is surrounded by 6 Na^+, and each Na^+ is surrounded by 6 Cl^-. The regular, repetitive structure has a high degree of order and low entropy or low disorder.

When solid NaCl is heated to 801°C, it melts, and the ions are no longer fixed in a simple geometric pattern. They will move relative to each other, subject only to the constraint of electrostatic attraction and repulsion. Each Na^+, therefore, will be adjacent to as many Cl^- anions as possible, and each Cl^- will tend to be surrounded by Na^+ cations. No longer would each ion be surrounded by precisely 6 ions of the opposite charge. This looser arrangement of ions in molten sodium chloride shows that the liquid state displays an increased disorder than the solid state, so it is of higher entropy.

A gas has even greater disorder than a liquid because its constituent molecules or atoms are no longer constrained to be adjacent to each other. Each gas particle moves more or less independently of the other particles. This state is one of near maximum disorder and near maximum entropy.

The second law of thermodynamics states that the total entropy of a chemical system *and* that of its surroundings always increases if the chemical or physical change is spontaneous. The preferred direction in nature is toward maximum entropy. Moving in the direction of greater disorder in an isolated system is one of the two forces that drive change. The other is loss of heat energy, ΔH.

Chemists have found it possible to assign a numerical quantity for the entropy of each substance. When measured at 25°C and 1 atm, these are called *standard entropies.* Table 14-4 lists 12 such values, symbolized by S° where the superscript denotes the standard state.

Table 14-4 Standard Entropies

Substance	$S^\circ\left(\frac{\text{joules}}{K-\text{mole}}\right)$	***Substance***	$S^\circ\left(\frac{\text{joules}}{K-\text{mole}}\right)$
CO (*g*)	198	N_2 (*g*)	192
CO_2 (*g*)	214	NO_2 (*g*)	240
H_2O (*l*)	70	N_2O_4 (*g*)	304
H_2O (*g*)	189	O_2 (*g*)	205
Li (*s*)	28	S (*s*)	32
Li (*g*)	138	SO_2 (*g*)	24.8

Notice in Table 14-4 that the units for entropy, $\frac{\text{joules}}{K-\text{mole}}$, require you to multiply each value by the temperature (*K*) in order to obtain units of energy.

The point that solids have low entropies and gases have high entropies has already been made. An examination of the values in the table should convince you that this is indeed a valid generalization. Compare the pairs of values for the two states of H_2O and also the two states of lithium.

The symbol for entropy is *S*, so a change in entropy is shown as ΔS. The values in the preceding chart allow calculations of the entropy change when water evaporates at 25°C.

$$H_2O(l) \rightarrow H_2O(g)$$

The entropy of reaction is the difference in the entropy of the products and reactants:

$$\Delta S = S_{\text{products}} - S_{\text{reactants}} = 189 - 70 = 119 \text{ J/deg}$$

This positive entropy change means that there is greater disorder in the product (H_2O gas) than in the reactant (H_2O liquid). In terms of just entropy, the increase in entropy drives the reaction to the right, toward a condition of higher entropy.

As another example, the entropy change associated with the reaction

$$2\ CO(g)+O_2(g)\rightarrow 2\ CO_2(g)$$

can be calculated from data in the chart of standard entropies:

$$\Delta S=\underbrace{[2(214)]}_{\text{product}}^{CO_2}-\underbrace{[2(\overset{CO}{198})+(\overset{O_2}{205})]}_{\text{reactants}}=-173\ \text{J/deg}$$

This negative entropy of reaction would tend to inhibit this reaction from proceeding.

The entropy of reaction by itself, however, is not sufficient to predict the direction of a reaction. At 25°C, you know that H_2O (*l*) is the stable phase, not H_2O (*g*). Moreover, the second reaction

$$2\ CO(g)+O_2(g)\rightarrow 2\ CO_2(g)$$

proceeds forward despite the negative entropy of reaction. You must consider both the enthalpy of reaction and the entropy of reaction in order to determine the direction of a chemical reaction with certainty.

Gibbs Free Energy

The American physicist Josiah Gibbs introduced (ca. 1875) a thermodynamic quantity combining enthalpy and entropy into a single value called free energy (or Gibbs free energy). In honor of its inventor, it is usually symbolized as *G*. The definition of free energy is

$$\underset{\text{free energy}}{G} = \underset{\text{enthalpy}}{H} - \underset{(\text{temperature})\ \text{entropy}}{TS}$$

where *T* is the temperature in Kelvin; entropies must be multiplied by temperature to get units of energy.

For studying chemical reactions, the relationship involves changes in the three thermodynamic quantities:

$$\underset{\substack{\text{change in}\\ \text{free energy}}}{\Delta G} = \underset{\substack{\text{change in}\\ \text{enthalpy}}}{\Delta H} - \underset{\substack{(\text{temperature})\ \text{change}\\ \text{in entropy}}}{T\Delta S}$$

A cardinal thermodynamic principle is that systems change toward minimum free energy. The sign of ΔG permits prediction of the behavior of a proposed chemical reaction with certainty.

Table 14-5 Sign of Gibbs Free Energy

ΔG	*Reaction Behavior*
Negative	Proceeds spontaneously to the right
Zero	Is at equilibrium
Positive	Will not proceed

Although the free energy change of a reaction can be calculated from the preceding equation, if ΔH and ΔS for the reaction are not known, it is much more common to use this alternative equation:

$$\Delta G_{\text{reaction}} = \sum \Delta G^{\circ}_{\text{f products}} - \sum \Delta G^{\circ}_{\text{f reactants}}$$

in which the values on the right are free energies of formation for each substance. For the standard state of 1 atm and 25°C, these values are called *standard free energies of formation.* As was the case with enthalpies of formation, the standard free energy of formation of an element is defined as zero. The value for each compound is the change in free energy associated with combining the elements to make one mole of the compound. In Table 14-6, the symbol ΔG°_{f} has a superscript denoting standard and a subscript denoting formation.

Table 14-6 Standard Free Energies of Formation (kJ/mole)

Compound	ΔG°_{f}	*Compound*	ΔG°_{f}
CO (*g*)	-137	NO (*g*)	87
CO_2 (*g*)	-394	NO_2 (*g*)	52
H_2 (*g*)	0.0	O_2 (*g*)	0.0
H_2O (*g*)	-229	S (*s*)	0.0
H_2O (*l*)	-237	SO_2 (*g*)	-300
H_2S (*g*)	-34	SO_3 (*g*)	-371

Given a listing of free energies of formation, the free energy change of a chemical reaction may be calculated in the same manner as you evaluated enthalpies of reaction and entropies of reaction. For the reaction that was discussed earlier,

$$2\,CO(g)+O_2(g)\rightarrow 2\,CO_2(g)$$

the free energy of reaction is calculated

$$\Delta G=\underbrace{\overset{CO_2}{[2(-394)]}}_{\text{product}}-\underbrace{[\overset{CO}{2(-137)}+\overset{O_2}{(0)}]}_{\text{reactants}}=-514\text{ kJ}$$

Since the change in free energy is negative, the forward reaction occurs spontaneously. Remember, systems change spontaneously to minimize free energy. Because the forward reaction loses free energy as it proceeds, that is the direction it will move spontaneously.

As another example of this important technique, consider the reaction involving oxides of both carbon and sulfur:

$$CO_2(g)+SO_2(g)\rightarrow CO(g)+SO_3(g)$$

From the free energies of formation given in Table 14-6, you can calculate the free energy of the reaction:

$$\Delta G=\underbrace{[\overset{CO}{(-137)}+\overset{SO_3}{(-371)}]}_{\text{products}}-\underbrace{[\overset{CO_2}{(-394)}+\overset{SO_2}{(-300)}]}_{\text{reactants}}=+186\text{ kJ}$$

This positive value for the free energy of reaction means that the reaction will not proceed to the right. Instead, the reverse reaction is favored, with CO and SO_3 reacting to yield CO_2 and SO_2. Moving to the left minimizes the free energy. Of course, this reaction can take place only if both carbon monoxide and sulfur trioxide are present initially. If you start with only carbon dioxide and sulfur dioxide, no reaction can occur.

The free energy calculations for both reactions

$$2\,CO(g)+O_2(g)\rightarrow 2\,CO_2(g)$$

and

$$CO_2(g)+SO_2(g)\rightarrow CO(g)+SO_3(g)$$

are valid only at 25°C and 1 atm because standard free energies of formation are utilized to calculate the free energy changes of the two reactions.

However, if you know both the standard enthalpies of formation and standard entropies for every substance in a reaction, you can estimate the free energy of reaction at other temperatures by using the equation

$$\Delta G = \Delta H - T\Delta S$$

which, you may recall, is Gibbs's definition of the free energy change.

Use the preceding equation to estimate the free energy change at 400°C for the reaction

$$2\ CO(g) + O_2(g) \rightarrow 2\ CO_2(g)$$

You need tabulated values of standard enthalpies of formation and standard entropies for the three gases, as shown in Table 14-7.

Table 14-7 Values at 25°C

Gas	ΔH_f° (kJ/mole)	S° (J/K mole)
CO (*g*)	−111	198
CO_2 (*g*)	−394	214
O_2 (*g*)	0	205

Using the values in Table 14-7, you should be able to calculate the changes for the reaction at 25°C and obtain

$$\Delta H = -566 \text{ kJ}$$

$$\Delta S = -173 \text{ J/deg}$$

Perform both calculations now to make sure that you understand how they were done. Notice that the energy units for the values differ by a factor of 1,000; in the calculations that follow, ΔS is divided by 1,000 to make the terms comparable, converting calories to kilocalories.

Use these values now to calculate the free energy of reaction from the equation:

$$\Delta G = \Delta H - T\Delta S$$

At 25°C, which equals 298 *K*,

$$\Delta G = (-566) - (298)(-0.173) = 514$$

At 400°C, which equals 673 *K*,

$$\Delta G = (-566) - (673)(-0.173) = 450 \text{ kJ}$$

The two results for the free energy of reaction differ because of the change in temperature. The first free energy change (at 25°C) is accurate because the tabulated values of enthalpies of formation and entropies are valid at that standard temperature. The second free energy change (at 400°C) is only a convenient estimate, based on the assumption that ΔH and ΔS do not change significantly with temperature. Actually, both do change slightly with temperature.

This final chapter has only begun to introduce you to the many thermodynamic calculations that assist chemists in understanding reactions. Table 14-8 provides information you should know about enthalpy, entropy, and free energy.

Table 14-8 Three Thermodynamic Quantities

Quantity	***Symbol***	***Measures***	***Units***
Enthalpy	*H*	Heat	Energy
Entropy	*S*	Disorder	Energy/K
Free energy	*G*	Reactivity	Energy

Don't forget that both the *calorie* and the *joule* are units of energy in published charts, so you will often have to do a conversion to obtain the unit that you want.

Problem 46: Use the chart of standard free energies of formation (Table 14-6) to determine the free energy change in the following reaction and report whether the reaction will proceed.

$$NO(g) + H_2O(l) \rightarrow NO_2(g) + H_2(g)$$

Problem 47: Use the standard values in Table 14-9 to compare the free energy change of the reaction at both 25°C and 100°C.

$$N_2O_4(g) \rightarrow 2\ NO_2(g)$$

Table 14-9 Data for Problem 47

Gas	*ΔH_f° (kJ/mole)*	*S° (cal/K mole)*
N_2O_4 (*g*)	9.661	304.3
NO_2 (*g*)	33.84	240.5

Chapter Check-Out

1. What is the sign of the enthalpy change, ΔH, for an exothermic reaction?

   a. positive
   b. negative
   c. No sign is needed.

2. Which process would be expected to have an increase in entropy?

   a. $Zn(s) \rightarrow Zn(l)$
   b. $Na^+(g) + Cl^-(g) \rightarrow NaCl(s)$
   c. $2\ H_2(g) + O_2(g) \rightarrow 2\ H_2O(l)$

3. Consulting the standard enthalpies of formation (Table 14-3), what is the enthalpy change, ΔH, for this reaction?

$$MgO(s) + H_2O(l) \rightarrow Mg(OH)_2(s)$$

   a. –924.7 kJ
   b. 1812.1 kJ
   c. 37.2 kJ

4. ΔG for the following reaction is calculated to be almost –69.5 kJ.

$$2\ NO(g) + O_2(g) \rightarrow 2\ NO_2(g)$$

   This means that a mixture of NO, O_2, and NO_2 will ______.

   a. react to form more NO_2 (a move to the right)
   b. react to form more NO and O_2 (a move to the left)
   c. not react in either direction

5. Which is the correct formulation for the relationship between free energy, enthalpy, and entropy?

a. $\Delta G = \Delta S + T\Delta H$

b. $\Delta G = \Delta H + T\Delta S$

c. $\Delta G = \Delta H - T\Delta S$

Answers: **1.** b **2.** a **3.** c **4.** a **5.** c

REVIEW QUESTIONS

Use this review to practice what you've learned in this book. After you work through the review questions, you're well on your way to understanding chemistry.

Chapter 1

1. Which of the following elements is similar to calcium (Ca)?

a. Ba
b. C
c. Na

2. Most the elements on the left side and middle of the periodic table are called ______.

a. noble gases
b. metals
c. nonmetals

3. The element with the atomic number 16 is ______.

a. O
b. S
c. H

Chapter 2

4. Which of the following represents one atom of carbon?

a. Ba
b. C
c. Na

5. What mass in grams represents one mole of carbon monoxide, CO?

a. 28 g
b. 14 g
c. 2 g

6. How many atoms are in one molecule of glucose $C_6H_{12}O_6$?

   a. 24

   b. 1

   c. 12

Chapter 3

7. How many protons are in the nucleus of an atom of iron?

   a. 26

   b. 56

   c. 23

8. Atoms of the same element that differ in the number of neutrons in the nucleus are called ______.

   a. ions

   b. isotopes

   c. alpha particles

9. Element X exists in nature in two isotopic forms: 60.0% of the atoms are isotope ^{30}X, which has an atomic weight of 29.88; 40.0% of the atoms are isotope ^{33}X, which has an atomic weight of 33.20. What is the weighted average atomic weight of element X?

   a. 31.21

   b. 31.54

   c. 33.41

Chapter 4

10. How many orbitals are in a *p*-type subshell?

    a. 1

    b. 5

    c. 3

11. All the elements in the left column of the periodic table, the alkali metals, have what number of valence electrons in what type of subshell?

    a. one valence electron in a *p*-subshell

    b. two valence electrons in an *s*-subshell

    c. one valence electron in an *s*-subshell

12. How many valence electrons are in one atom of phosphorus, one atom of sulfur, and one atom of bromine, respectively?

a. 3, 2, and 1
b. 5, 6, and 7
c. 3, 2, and 7

Chapter 5

13. Considering the number of valence electrons in oxygen (O) and fluorine (F), what is the expected formula of a compound formed between these two elements?

a. OF
b. O_2F
c. OF_2

14. Considering the number of valence electrons in aluminum (Al) and oxygen (O), what is the expected formula for a compound formed from these two elements?

a. Al_2O_3
b. AlO_2
c. Al_2O

15. The Lewis structure for carbon dioxide, CO_2, begins with two oxygen atoms joined to carbon: O-C-O. What kinds of covalent bonds connect carbon with the two oxygen atoms in carbon dioxide, CO_2?

a. two single bonds
b. two double bonds
c. one single bond and one double bond

Chapter 6

16. Which of the following represents a hydrocarbon?

a. C_2H_5OH
b. CH_2Cl_2
c. C_3H_8

17. The fact that there are five known organic compounds with the same formula C_6H_{14} demonstrates the existence of _______.

- **a.** hydrocarbons
- **b.** isomers
- **c.** hexane

18. An organic compound with a formula of C_6H_{14} is most likely an _______.

- **a.** alkane
- **b.** alkene
- **c.** alkyne

Chapter 7

19. Which of the following phase changes releases energy?

- **a.** solid → liquid
- **b.** solid → gas
- **c.** gas → liquid

20. It requires 75.40 joules to raise the temperature of 1.0 mole of liquid water 1.0°C. How many calories of heat would be required to warm 5.0 moles of water from 15°C to 60°C?

- **a.** 16,966 joules
- **b.** 3,397 joules
- **c.** 377 joules

21. In which phase change are the forces between molecules being completely broken down?

- **a.** liquid → gas
- **b.** solid → liquid
- **c.** gas → liquid

Chapter 8

22. What is the mass of Avogadro's number of water molecules?

- **a.** 18.0 grams
- **b.** 10.0 grams
- **c.** 6.02×10^{23} grams

23. What is the final volume of 20.0 L of air if the pressure exerted on the air is doubled while its absolute temperature doubles?

a. 20.0 L

b. 10.0 L

c. 40.0 L

24. Consider the reaction of hydrogen and nitrogen to form ammonia:

$$3\ H_2 + N_2(g) \rightarrow 2\ NH_3(g)$$

How many liters of ammonia will form for each 5 liters of nitrogen consumed?

a. 5 L

b. 10 L

c. 2 L

Chapter 9

25. What is the molarity of glucose, $C_6H_{12}O_6$, in a solution that contains 4.50 g of glucose in 0.50 L of solution? The molecular mass of glucose is 180.

a. 0.050 M

b. 0.025 M

c. 1.0 M

26. Silver chloride, AgCl, is of very low solubility in water. A saturated solution of AgCl is 1.3×10^{-5} M at room temperature. What is the K_{sp} of AgCl?

a. 1.3×10^{-5}

b. 2.6×10^{-5}

c. 1.7×10^{-10}

27. What is the molarity of ions in a 0.25 M solution of $AlCl_3$?

a. 0.25 M

b. 0.75 M

c. 1.0 M

Chapter 10

28. Which set of compounds is a strong acid, a weak acid, and a base, in that order?

a. CH_3COOH, NH_3, HCl

b. HNO_3, H_2CO_3, $Ca(OH)_2$

c. HCl, H_2SO_4, CH_3COOH

29. Which acid will have a hydrogen ion concentration less than the concentration of the acid itself in water?

a. H_2SO_3

b. HNO_3

c. HCl

30. What is the pH of a solution that has a hydrogen ion concentration of 2×10^{-9} M?

a. 9.0

b. 8.7

c. 2.0

Chapter 11

31. Which element is oxidized and which is reduced in this equation?

$$O_2(g) + 4\,H^+(aq) + 4\,Fe^{2+}(aq) \rightarrow 4\,Fe^{3+}(aq) + 2\,H_2O(l)$$

a. Fe^{2+} is oxidized; O_2 is reduced.

b. H^+ is oxidized; Fe^{2+} is reduced.

c. O_2 is oxidized; Fe^{2+} is reduced.

32. What is the oxidation number of Cr in $Cr_2O_7^{2-}$?

a. +3

b. +6

c. +7

33. What is the coefficient placed in front of NO_3^- when this redox equation is correctly balanced?

$$NO_3^-(aq) + H^+(aq) + Sn(s) \rightarrow NO_2(g) + H_2O(l) + SN^{2+}(aq)$$

a. 1

b. 2

c. 3

Chapter 12

34. The following reaction can be made to occur in a voltaic cell:

$$2\ I^-(aq) + Pt^{2+}(aq) \rightarrow I_2(s) + Pt(s)$$

Which reaction occurs at the cathode?

a. $2\ I^-(aq) + Pt^{2+}(aq)$
b. $2\ I^-(aq) \rightarrow I_2(s)$
c. $Pt^{2+}(aq) \rightarrow Pt(s)$

35. Using the data in Table 12-1, what is the expected voltage of a voltaic cell using this reaction?

$$3\ Ag^+(aq) + Al(s) \rightarrow 3\ Ag(s) + Al^{3+}(aq)$$

a. +2.46 v

b. +0.86 v

c. –0.86 v

36. The electrolysis of molten sodium chloride produces pure sodium metal (Na) and chlorine gas (Cl_2). How many faradays of electricity are needed to produce 500 grams of Cl_2? The half-reaction forming Cl_2 is

$$2\ Cl^-(l) \rightarrow Cl_2(g) + 2\ e^-$$

a. 7.0 faradays

b. 14 faradays

c. 3.5 faradays

Chapter 13

37. Hydrogen iodide (HI) decomposes into iodine (I_2) and hydrogen (H_2) at 300°C and establishes the following equilibrium:

$$\text{Heat} + 2\ \text{HI}(g) \rightleftarrows \text{H}_2(g) + \text{I}_2(g)$$

Which of the following will shift the equilibrium back to the left?

a. increasing the concentration of HI

b. increasing the temperature

c. increasing the concentration of I_2

38. What is the value of the equilibrium constant, K, for the reaction

$$2\ \text{HI}(g) \rightleftarrows \text{H}_2(g) + \text{I}_2(g)$$

if at equilibrium the concentrations are the following:

$$[\text{HI}] = 0.62\ \text{M},\ [\text{H}_2] = 0.089\ \text{M, and } [\text{I}_2] = 0.089\ \text{M?}$$

a. 0.021

b. 0.013

c. 0.46

39. Again, considering the same reaction:

$$\text{Heat} + 2\ \text{HI}(g) \rightleftarrows \text{H}_2(g) + \text{I}_2(g)$$

The reverse reaction is:

a. neutral

b. endothermic

c. exothermic

Chapter 14

40. At 298 K, the enthalpy change of a reaction is –120 kJ and the entropy change is +0.131 kJ/K. What is the free energy change for the reaction?

a. –159 kJ

b. –81 kJ

c. +11 kJ

41. Which change represents an increase in entropy?

a. $CO_2(g)+H_2O(l)\rightarrow H_2CO_3(aq)$
b. $CaCO_3(s)\rightarrow CaO(s)+CO_2(g)$
c. $HCl(g)\rightarrow HCl(l)$

42. Using the data in the Table 14-6, the Standard Free Energies of Formation, calculate the change in free energy, (Δ, for this reaction):

$$2\ SO_2(g)+O_2(g)\rightarrow 2\ SO_3(g)$$

a. –33.4 kJ
b. +16.7 kJ
c. –320.6 kJ

Answers: **1.** a. **2.** b. **3.** b. **4.** b. **5.** a. **6.** a. **7.** a. **8.** b. **9.** a. **10.** c. **11.** c. **12.** b. **13.** c. **14.** a. **15.** b. **16.** c. **17.** b. **18.** a. **19.** c. **20.** a. **21.** a. **22.** a. **23.** a. **24.** b. **25.** a. **26.** c. **27.** c. **28.** b. **29.** a. **30.** b. **31.** a. **32.** b. **33.** b. **34.** c. **35.** a. **36.** b. **37.** c. **38.** a. **39.** c. **40.** a. **41.** b. **42.** a.

RESOURCE CENTER

The Resource Center offers the best resources available in print and online to help you study and review the core concepts of chemistry. You can find additional resources, plus study tips and tools to help test your knowledge, at www.cliffsnotes.com.

Books

This CliffsNotes book is one of many great books about chemistry. If you are interested in additional resources, we suggest the following publications:

CliffsNotes Chemistry Practice Pack includes a CD-ROM that includes many practice questions for you to continue testing yourself. John Wiley and Sons, $18.99.

Chemistry For Dummies covers the basics in an entertaining way to learn. John Wiley and Sons, $19.99.

CliffsNotes AP Chemistry with CD-ROM, 4th Edition, is perfect for students taking the AP Chemistry exam and includes timed and untimed exams on the CD-ROM. John Wiley and Sons, $29.99.

Wiley also has three Web sites that you can visit to read about all the books we publish:

- www.cliffsnotes.com
- www.dummies.com
- www.wiley.com

Internet

Visit the following Web sites for more information about chemistry:

General Chemistry Online!, http://antoine.frostburg.edu/chem/senese/101/index.shtml. A rich source of chemistry facts with links to many areas of chemistry including the latest developments. Click

General Chemistry for tutorials and frequently asked questions, and you can even ask your own questions as well as search for a particular topic.

About Chemistry, http://chemistry.about.com/science/chemistry. An open door to the basics of chemistry and beyond with links to tutorials, all the areas of chemistry, the history of chemistry, and latest discoveries. Click General Chemistry to learn the basics.

Chemistry Help, http://chemistryhelp.net. Includes chemistry games and projects that reinforce chemistry basics.

Chemistry Tutor, http://library.thinkquest.org/2923/. Gets you through chemistry.

Chemistry Team, http://www.chemteam.info/ChemTeamIndex.html. A tutorial for high school chemistry with links to individual chemistry topics that were discussed in this book.

GLOSSARY

acid a compound that yields H^+ ions in solution or a solution in which the concentration of H^+ exceeds OH^-.

acid ionization constant the equilibrium constant describing the degree of ionization of an acid.

actinides the row of elements below the main section of the periodic table, from thorium to lawrencium.

alkali synonym for base.

alkali metals the column of elements from lithium to francium.

alkaline earths the column of elements from beryllium to radium.

alkane a hydrocarbon without a double bond, triple bond, or ring structure.

alkene a hydrocarbon with one or more double bonds and no triple bond.

alkyne a hydrocarbon with one or more triple bonds.

alpha particle a cluster of two protons and two neutrons emitted from a nucleus in one type of radioactivity.

anion a chemical species with a negative charge.

anode the electrode at which oxidation occurs.

aqueous refers to a solution with water as the solvent.

aromatic refers to an organic compound with a benzene-like ring.

atom the smallest amount of an element; a nucleus surrounded by electrons.

atomic number the number of protons in the nucleus of the chemical element.

atomic mass the mass in grams of one mole of the chemical element; approximately the number of protons and neutrons in the nucleus.

Avogadro's law equal volumes of gases at the same temperature and pressure contain the same number of molecules.

Avogadro's number 6.02×10^{23}, the number of molecules in one mole of a substance.

base a compound that yields OH^- ions in solution or a solution in which the concentration of OH^- exceeds H^+.

beta particle an extremely fast moving electron emitted from a nucleus in one type of radioactivity.

boiling point the temperature at which a liquid changes to a gas. (Technically, boiling point is the temperature at which the vapor pressure of the liquid is equal to atmospheric pressure.)

boiling point elevation an increase in the boiling point of a solution, proportional to the concentration of solute particles.

Boyle's law the volume of a gas varies inversely with pressure.

calorie a unit of energy, equal to 4.184 joules.

catalyst a substance that accelerates a chemical reaction without itself being consumed.

cathode the electrode at which reduction occurs.

cation an atom or molecule with a positive charge.

Charles' law the volume of a gas varies directly with absolute temperature.

chemical equation a shorthand way of describing a chemical change using symbols of elements and formulas of compounds.

chemical formula a representation of a species to show its composition using symbols, subscript numbers, and if necessary superscript numbers.

compound a substance formed by the chemical combination of two or more elements.

concentration the relative amount of a solute in a solution.

congeners elements with similar properties, arranged in columns of the periodic table.

conjugate an acid and base that are related by removing or adding a single hydrogen ion.

covalent bond atoms linked together by sharing valence electrons.

critical point a point in a phase diagram where the liquid and gas states cease to be distinct.

crystalline the regular, geometric arrangement of atoms in a solid.

decomposition a chemical reaction in which a compound is broken down into simpler compounds or elements.

dissociation the separation of a solute into constituent ions.

electrochemical cell a device that uses a chemical reaction to produce or use an electric current.

electrode the point in an electrochemical cell at which reduction or oxidation occurs.

electrolysis the decomposition of a substance by an electric current.

electrolyte a substance that forms ions when dissolved in water.

electromotive force the electrical potential produced by a chemical reaction, voltage.

electron a light subatomic particle with negative charge; found in orbitals surrounding an atomic nucleus.

electronegativity a number describing the attraction of an element for electrons in a chemical bond.

element a substance that cannot be decomposed; each chemical element is characterized by the number of protons in the nucleus.

EMF *See* electromotive force.

endothermic refers to a reaction that absorbs requires heat.

energy the concept of motion or heat required to do work.

enthalpy the thermodynamic quantity measuring the heat content of a substance.

entropy the thermodynamic quantity measuring the disorder of a substance.

equilibrium a balanced condition resulting from two opposing reactions or processes.

equilibrium constant the ratio of concentrations of products to reactants for a reaction at chemical equilibrium.

exothermic refers to a reaction that releases heat.

faraday a unit of electric charge equal to that on one mole of electrons.

Faraday's laws two laws of electrolysis relating the amount of substance to the quantity of electric charge.

fluid any substance that flows or deforms under an applied shear stress; any liquid or gas.

free energy the thermodynamic quantity measuring the tendency of a reaction to proceed; also called Gibbs free energy.

freezing point the temperature at which a liquid changes to a solid.

freezing point depression the decrease in freezing point of a solution, proportional to the concentration of solute particles.

fusion melting.

gamma radiation energy released from a nucleus during radioactive decay.

gas a state of matter in which molecules are widely separated, fluid, expandable, and compressible.

gas constant R equals 0.082 liter—atmospheres per mole—unit Kelvin.

gram formula mass an amount of a substance equal in grams to the sum of the atomic mass.

ground state the electronic configuration of lowest energy for an atom.

group a column of elements in the periodic table.

half-reaction an oxidation or reduction reaction with free electrons as a product or reactant.

halogens the column of elements from fluorine to astatine.

heat a form of energy that spontaneously flows from a warm body to a cold body.

heat capacity the amount of energy needed to raise the temperature of a substance by one degree Celsius.

hydrocarbon an organic compound containing only carbon and hydrogen.

hydrogen bond a weak, secondary bond between a partially positive hydrogen atom and a partially negative N, O, or F atom; an intermolecular force of attraction.

hydroxide refers to the OH^- ion.

ideal gas equation the equation relating the volume of a gas to its pressure, temperature, and moles of gas.

inert gases the column of elements from helium to radon; also called noble gases.

ion an atom with an electric charge due to gain or loss of electrons.

ionic bond atoms linked together by the attraction of opposite charges.

ionic dissociation separation of cations and anions from an ionic compound by polar water molecules.

ionization removing electrons from an atom; alternatively, the dissociation of a solute into ions.

isoelectronic refers to several dissimilar atoms or ions with identical electronic configurations.

isomers several molecules with the same composition but different structures.

isotope a variety of an element characterized by a specific number of neutrons in the nucleus.

joule a unit of energy equal to 0.239 calorie.

lanthanides the row of elements beneath the main section of the periodic table, from cerium to lutetium; also called rare earths.

Le Chatelier's principle a system that in equilibrium is disturbed adjusts so as to minimize the disturbance.

liquid a state of matter in which the molecules are touching, fluid, incompressible.

litmus an indicator that turns red in acid and blue in alkaline solution.

mass number the sum of the number of neutrons and protons in the nucleus of an atom.

mass percent the percentage found by taking the mass of an element and dividing it by the mass of a sample and then multiplying by 100.

melting point the temperature at which a solid changes to a liquid.

metallic bond atoms linked together by the migration of electrons from atom to atom.

metals the elements in the middle and left parts of the periodic table, except for hydrogen.

molality the number of moles of solute in one kilogram of solvent.

molar heat capacity the amount of heat required to raise the temperature of one mole of substance one degree Celsius.

molarity the number of moles of solute in one liter of solution.

mole an amount of a substance equal in grams to the sum of the atomic weights.

mole fraction the fraction of moles (or molecules) of one substance in the total moles (or molecules) of all substances in the mixture. If the mole fraction of substance A is 0.1, one-tenth of all the molecules in a mixture are A molecules.

molecular formula describes the ratio of moles of the elements in a molecule.

molecule a group of atoms linked together by covalent bonds.

neutralization the chemical reaction of an acid and base to yield a salt and water.

neutron a heavy subatomic particle with zero charge; found in an atomic nucleus.

noble gases the column of elements from helium to radon; also called inert gases.

nonmetals the elements in the upper right part of the periodic table, and also hydrogen.

nucleon a proton or neutron found in an atomic nucleus.

nucleus the core of an atom, containing protons and neutrons.

orbital a region in space surrounding the nucleus, which is occupied by up to two electrons.

organic refers to compounds based on carbon.

organic chemistry an area of chemistry dealing principally with the chemistry of carbon.

oxidation the loss of electrons by a species.

oxidation number a signed integer representing the real or hypothetical charge on an atom due to the gain or loss of electrons.

oxidation-reduction reaction reaction in which electrons are transferred between atoms.

oxide a compound of oxygen and another element.

oxidizing agent the species that is reduced during a redox reaction.

period a horizontal row of elements in the periodic table.

periodic table display of the elements in order of atomic number with similar elements falling into columns.

pH a number describing the concentration of hydrogen ions in a solution; equals $-\log[H^+]$.

phase a substance with uniform composition and definite physical state.

polar bond a bond with both ionic and covalent characteristics.

polyprotic refers to an acid with several hydrogens that can ionize.

precipitate a solid that separates from solution.

product a substance on the right side of a chemical equation, which is produced or formed during a chemical reaction.

proton a heavy subatomic particle with a positive charge; found in an atomic nucleus.

radioactive decay the spontaneous disintegration of an unstable nucleus producing a different nucleus and various types of radiation particles.

radioactivity the emission of subatomic particles from a nucleus.

rare earths the elements from cerium to lutetium; lanthanides.

reactant a substance on the left side of a chemical equation which is one of the beginning substances in a chemical reaction.

redox refers to a reaction in which simultaneous reduction and oxidation occur.

reducing agent the species that is oxidized during a redox reaction.

reduction the gain of electrons by a species.

salt a solid compound composed of both metallic and nonmetallic elements, often as ions.

saturated describes a solution that holds as much solute as possible.

saturated hydrocarbon a hydrocarbon containing only single bonds between the carbon atoms

shell a set of electron orbitals with the same principal quantum number.

solid a state of matter in which the molecules are touching and possessing rigid shape and which is not compressible.

solubility the upper limit of concentration of a solute.

solubility product the constant obtained by multiplying the ion concentrations in a saturated solution raised to their balancing coefficients.

solute the substance that is dissolved in a solution.

solution a homogeneous mixture consisting of a solvent and one or more solutes.

solvent the host substance of dominant abundance in a solution.

specific heat capacity the amount of heat required to raise 1 gram of a substance 1°C.

standard temperature and pressure 0°C and 1 atmosphere.

states of matter solid, liquid, and gas.

stoichiometric refers to compounds or reactions in which the components are in fixed, whole-number ratios.

STP *See* standard temperature and pressure.

strong electrolyte an acid, base, or salt that dissociates almost completely to ions in aqueous solution.

structural formula depicts the bonding of atoms in a molecule.

sublimation the transformation of a solid directly to a gas without an intervening liquid state.

subshell a set of electron orbitals with the same principal and second quantum number; for example, $2p$ (subshell), $3s$ (subshell), and so on.

supersaturated describes a solution that holds more solute than is theoretically possible at that given temperature.

symbol an abbreviation for the name of an element; for example, C for carbon.

thermodynamics the study of the energy, principally heat energy, that accompanies chemical or physical changes.

transition metals the three rows of elements in the middle of the periodic table, from scandium to zinc, yttrium to cadmium. and lanthanum to mercury.

transmutation the process of changing one chemical element to another element during radioactive decay.

triple point a point in a phase diagram where the three states of matter are in equilibrium.

unsaturated describes a solution that does not hold as much solute as possible.

unsaturated hydrocarbon a hydrocarbon that contains double or triple bonds between at least two of its carbon atoms

valence a signed integer describing the combining power of an atom.

valence electrons the outermost shell of electrons in an atom or ion.

voltaic cell a device that uses a chemical reaction to produce electricity.

weak electrolyte an acid, base, or salt that dissociates only slightly to form ions in solution.

APPENDIX

Following are answer explanations to the Problems located throughout the chapters. Hopefully, you've challenged yourself to attempt some of these as you've reviewed the various topics. If not, you can use these problems as additional review now.

1. By mass, the compound is 26.19% nitrogen, 7.55% hydrogen, and 66.26% chlorine. (Total mass: 53.50 g)

$$\text{Nitrogen} \quad 1 \times 14.01 = 14.01 / 53.50 = 0.2619$$
$$\text{Hydrogen} \quad 4 \times 1.01 = 4.04 / 53.50 = 0.0755$$
$$\text{Chlorine} \quad 1 \times 35.45 = 35.45 / 53.50 = 0.6626$$

2. The simplest formula is K_2CuF_4.

$$\text{Potassium} \quad \frac{35.91}{39.10} = 0.918 / 0.459 = 2$$
$$\text{Copper} \quad \frac{29.19}{63.55} = 0.459 / 0.459 = 1$$
$$\text{Flourine} \quad \frac{34.90}{19.00} = 1.837 / 0.459 = 4$$

3. There are 6.37 moles of C_6H_5Br.

$$1 \text{ mole} = 6(12.01) + 5(1.01) + 1(79.90) = 157.01 \text{ grams/mole}$$
$$\frac{1000 \text{ g } C_6H_5Br}{157.01 \text{ g/mole}} = 6.37 \text{ moles}$$

4. The neon has a mass of 4.5 grams.

$$\frac{5 \text{ liters Ne}}{22.4 \text{ g/mole}} = 0.233 \text{ moles} \times 20.18 \text{ g/mole} = 4.5 \text{ grams}$$

5. The reaction uses 2.79 liters of oxygen.

$$1 \text{ mole } CH_4 = 1(12.01) + 4(1.01) = 16.05 \text{ grams}$$

$$\frac{1 \text{ gram}}{16.05 \text{ g/mole}} = 0.0623 \text{ mole} \times 22.4 \frac{\text{L}}{\text{mole}} = 1.395 \text{ liters } CH_4$$

The reaction coefficients denote the relative volumes, so the volume of O_2 is twice that of CH_4 = 2 × 1.395 L = 2.79 liters.

6. Nuclei B and C are isotopes of magnesium, the element with atomic number 12. Nuclei A and B both have a mass of approximately 24 atomic mass units because their nucleons sum to 24.

7. Natural silver is 48.15% silver-109.

$$x = \text{fraction } {}^{109}\text{Ag and } (1-x) = \text{fraction } {}^{107}\text{Ag}$$

$$108.905(x) + 106.905(1-x) = 107.868$$

$$2x + 106.905 = 107.868$$

$$x = 0.4185$$

8. The nucleus is radium-226, also written ${}^{226}_{88}\text{Ra}$. The atomic mass decreased by four, the mass of the alpha particle. The atomic number decreased by two because the alpha particle carried off two protons. The problem asks for the mass number "226," the atomic number "88," and the name of the element "radium."

9. Aluminum has three valence electrons; whereas oxygen has six. Remember that you count columns from the left margin of the periodic table.

10. The Lewis diagram for H_2S is

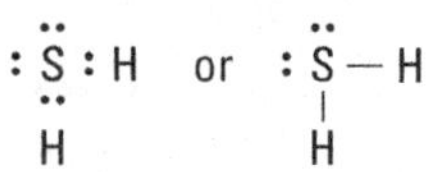

11. The electronegativity difference of magnesium and chlorine is 1.8,

$$\underset{\text{Cl}}{3.0} - \underset{\text{Mg}}{1.2} = 1.8$$

which corresponds to a bond with 52% ionic character and 48% covalent character. Such an intermediate bond is called polar.

12. The three isomers of C_5H_{12} are shown in the following example. The essential feature is the bonding of carbons. In the first molecule, no carbon is bonded to more than two carbons, the second molecule has a carbon bonded to three carbons, and the third molecule has a carbon bonded to four carbons.

n-pentane $CH_3-CH_2-CH_2-CH_2-CH_3$

isopentane

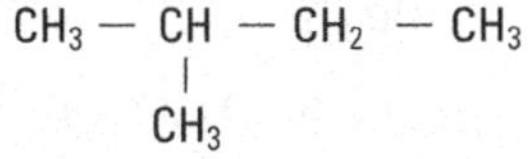

neopentane

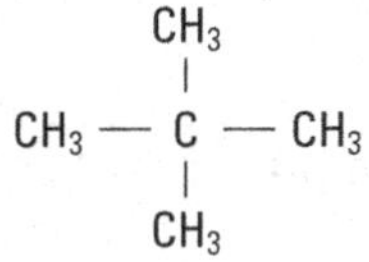

13. The addition of hydrogen converts acetylene to ethane:

$$\underset{\text{acetylene}}{C_2H_2}+\underset{\text{hydrogen}}{2\,H_2}\rightarrow\underset{\text{ethane}}{C_2H_6}$$

Because the number of moles of hydrogen is twice that of acetylene, the reaction requires 200 liters of hydrogen, double that of acetylene.

14. It is an aldehyde with the structural formula:

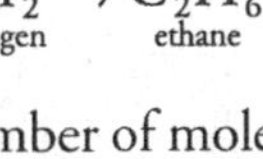

formaldehyde

15. The minimum pressure for liquid CO_2 is 5.1 atmospheres.

16. At –64°C, the solid CO_2 sublimates to the gas state.

17. The total heat needed is 49,831 joules.

$$5.55\text{ moles}\times 40°C\times 36.57\frac{J}{°C\text{-mole}}=811.8\text{ J}$$

(warming ice to 0°)

$$5.55\text{ moles}\times 6{,}008\frac{J}{\text{mole}}=33{,}344\text{ J (heat of fusion)}$$

$$5.55\text{ moles}\times 20°C\times 75.40\frac{J}{°C\text{-mole}}=8{,}369\text{ J}$$

(warming water to 20°)

18. The pressure equals 0.804 atmosphere.

$$611 \text{ mm Hg} \times \frac{1 \text{ atm}}{760 \text{ mm Hg}} = 0.804 \text{ atm}$$

19. The required pressure is 1.654 atmospheres.

$$(1.13 \text{ atm})(732 \text{ ml}) = (x \text{ atm})(500 \text{ ml})$$

$$x = (1.13 \text{ atm})\frac{732 \text{ ml}}{500 \text{ ml}} = 1.654 \text{ atm}$$

20. The chilled temperature is –217.63°C.

$$\frac{\text{V}}{\text{T}} = \frac{660 \text{ ml}}{293.15 \text{ K}} = \frac{125 \text{ ml}}{x}$$

$$x = (293.15 \text{ K})\frac{125 \text{ ml}}{660 \text{ ml}} = 55.52 \text{ K} = -217.63°\text{C}$$

21. There are 1.5×10^{24} hydrogen atoms.

$$\frac{10 \text{ grams}}{16.05 \text{ g/mole}} = 0.623 \text{ mole CH}_4$$

$$0.623 \text{ mole CH}_4\left(\frac{4 \text{ moles H}}{1 \text{ mole CH}_4}\right)\left(\frac{6.02 \times 10^{23} \text{ H atoms}}{1 \text{ mole H}}\right)$$

$$= 1.5 \times 10^{24} \text{ H atoms}$$

22. The carbon monoxide occupies 28,499 liters.

$$\frac{1{,}000 \text{ grams}}{28.01 \text{ g/mole}} = 35.70 \text{ moles CO}$$

$$\text{V} = \frac{\text{nRT}}{\text{P}} = \frac{(35.70)(0.082)(973.15)}{(0.1)} = 28{,}499 \text{ liters}$$

23. The ozone molecule has the formula O_3.

$$\frac{1.073 \text{ liters}}{22.4 \text{ liters/mole}} = 0.0479 \text{ mole}$$

$$\frac{2.3 \text{ grams}}{0.0479 \text{ mole}} = 48 \text{ grams/mole (molecular weight)}$$

$$\frac{48 \text{ g(molecular weight)}}{16 \text{ g per oxygen atom}} = 3 \text{ oxygen atoms per O}_3 \text{ molecule}$$

24. The solution is 0.592 in glucose.

$$C_6H_{12}O_6 = 6(12.01) + 12(1.01) + 6(16.00) = 180.18 \text{ grams/mole}$$

$$\text{moles glucose} = \frac{80 \text{ g}}{180.18 \text{ g/mole}} = 0.444 \text{ mole}$$

$$\text{molality} = \frac{0.444 \text{ mole glucose}}{0.750 \text{ Kg solvent}} = 0.592 \text{ m}$$

25. The solution is 0.36 mole fraction alcohol.

CH_3OH = 32.05 grams/mole

H_2O = 18.02 grams/mole

moles alcohol = 100 g/32.05 g/mole = 3.12 mole

moles water = 100 g/18.02 g/mole = 5.55 mole

$$\text{mole fraction alcohol} = \frac{(3.12)}{(3.12 + 5.55)} = 0.36$$

26. The amount of CuCl is 0.00152 mole. If the powder dissolved completely, the solution would be 0.00152 molar with respect to both Cu^+ and Cl^-.

$[Cu^+]\ [Cl^-] = (0.00152)^2 = 2.3 \times 10^{-6}$

Because that product exceeds the solubility product given in the chart as 1.1×10^{-6}, which is the value for a saturated solution, the powder will not completely dissolve.

27. The solubility of aluminum hydroxide is 0.00843 gram per liter. The $Al(OH)_3$ dissociates to 4 ions with the concentration of OH^- being three times that of A^{3+}.

$$[Al^{3+}][OH^-]^3 = K_{sp}$$

$$(x)(3x)^3 = 1.9 \times 10^{-33}$$

$$27x^4 = 1.9 \times 10^{-33}$$

$$x^4 = 7.04 \times 10^{-35}$$

$$x = 2.90 \times 10^{-9} = \text{ molarity of } Al(OH)_3$$

$(1.08 \times 10^{-4} \text{ mole})(78.01 \text{ g/mole}) = 2.25 \times 10^{-7}$

28. The sodium chloride solution boils at 100.87°C.

$$\frac{10 \text{ g NaCl}}{58.44 \text{ g/mole}} = 0.171 \text{ mole}$$

$$\frac{0.171 \text{ mole}}{0.2 \text{ kg } H_2O} = 0.855 \text{ molal NaCl}$$

Each formula unit yields 2 ions; so, the total molality of the ions is twice that, or 1.712 m. The change in boiling point is

$$\Delta T_b = 1.712 \text{ m}\left(\frac{0.51°C}{m}\right) = 0.87°C$$

and that value is added to the 100° boiling point of pure water.

29. The molecular weight of brucine is approximately 394. Table 9-3 states that pure chloroform freezes at –63.5°C.

$$\Delta T_f = (-63.5°C) - (-64.67°C) = 1.19°C$$

$$m = \frac{\Delta T_f}{K_f} = \frac{1.19°C}{4.68°C/m} = 0.254 \text{ molal} = 0.254 \text{ mole/kg}$$

$$\frac{100 \text{ g/kg}}{0.254 \text{ mole/kg}} = 394 \text{ grams/mole}$$

30. The solution is alkaline with pH = 8.34.

$$\left[H^+\right] = \frac{K_W}{\left[OH^-\right]} = \frac{1 \times 10^{-14}}{2.2 \times 10^{-6}} = 4.55 \times 10^{-9} \text{ m}$$

$$pH = -\log_{10}\left(4.55 \times 10^{-9}\right) = 8.34 (\text{over 7, so alkaline})$$

31. The solution required 0.056 mole of acetic acid.
From the pH, $[H^+] = 10^{-3}$ and $[CH_3COO^-]$ must be the same.

32. The conjugate base of HCO_3^- is the carbonate ion CO_3^{2-}, formed by the loss of a proton. The conjugate acid is carbonic acid H_2CO_3, formed as HCO_3^- gains a proton.

33. $\left[H_2SO_3\right] > \left[H^+\right] > \left[HSO_3^-\right] \gg \left[SO_3^{2-}\right]$

34. Nitrogen has the oxidation number –3 in Mg_3N_2 and +5 in HNO_3. For Mg_3N_2,

$$\overset{\text{Mg}}{3(+2)}+\overset{\text{N}}{2(x)}=0,\ \text{so } x=-3$$

For HNO_3,

$$\overset{\text{H}}{1(+1)}+\overset{\text{N}}{1(x)}+\overset{\text{O}}{3(-2)}=0,\ \text{so } x=+5$$

Notice that the oxidation number per atom is multiplied by the number of atoms in the formula unit.

35. Carbon is oxidized and iodine is reduced, so CO is the reducing agent and I_2O_5 is the oxidizing agent.

$$I_2^{(+5)}O_5^{(-2)}+5C^{(+2)}O^{(-2)}\rightarrow I_2^{(0)}+5C^{(+4)}O_2^{(-2)}$$

Each of the five carbon atoms loses two electrons, and each of the two iodine atoms gains five electrons.

36. Only manganese and oxygen have variable oxidation numbers.

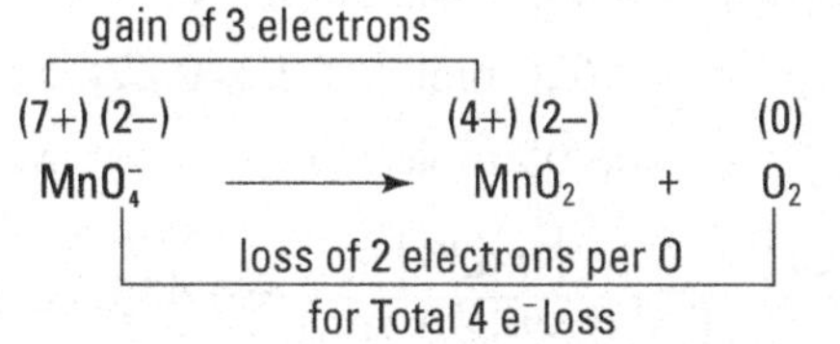

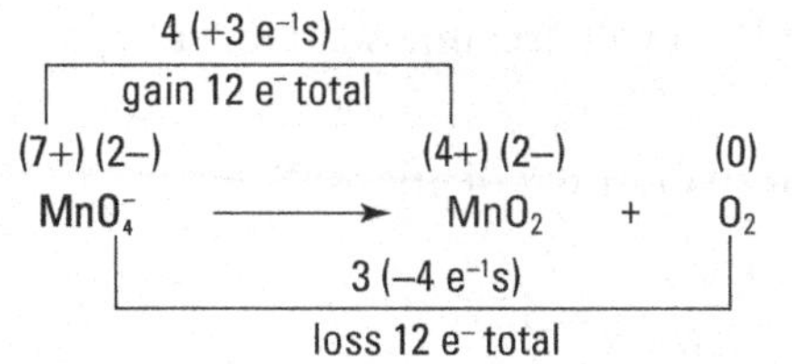

$$4MnO_4^- + 4MnO_2 + 3O_2^0 + 2H_2O$$

37. The silver is deposited from solution as the iron dissolves.

oxidation	$Fe(s)\rightarrow Fe^{2}+(aq)+2e^-$
reduction	$2Ag^+(aq)+2e^-\rightarrow 2Ag(s)$
overall	$Fe(s)+2Ag^+(aq)\rightarrow 2Ag(s)+Fe^{2}+(aq)$

38. The lithium-fluorine battery yields 5.91 volts.

Half-Reaction	*Type*	*Potential*
$2Li \rightarrow 2Li^+ + 2e^-$	oxidation	3.04 v
$F_2 + 2e^- \rightarrow 2F^-$	reduction	2.87 v

39. The electrolysis requires 111.2 faradays of electricity.

$$\frac{1{,}000 \text{ grams Al}}{26.98 \text{ g/mole}} = 37.06 \text{ moles Al}$$

$Al^{(+3)} + 3e^- \rightarrow Al^{(0)}$ (reduction)

moles of electrons = 3 × moles Al = 3 × 37.06 = 111.2 moles e^-

40. The initial reaction is: $Br_2(g) + Cl_2(g) \rightleftarrows 2\ BrCl(g)$

$$\text{moles BrCl} = 50/115 = 0.433 \text{ mole}$$
$$\text{moles } Br_2 = 100/159.8 = 0.626 \text{ mole}$$
$$\text{moles } Cl_2 = 50/70.9 = 0.705 \text{ mole}$$
$$\text{molarity BrCl} = 0.433 \text{ mole/1.0 L} = 0.433 \text{ M}$$
$$\text{molarity } Br_2 = 0.626 \text{ mole/1.0 L} = 0.626 \text{ M}$$
$$\text{molarity } Cl_2 = 0.705 \text{ mole/1.0 L} = 0.705 \text{ M}$$
$$\frac{[BrCl]^2}{[Br_2][Cl_2]} = \frac{[0.433]^2}{[0.626][0.705]} = 0.425 > 0.15$$

Conclusion: BrCl will decompose to form Br_2 and Cl_2 to restore the equilibrium.

41. The value of PSO_3 is 0.274 atmosphere.

$$\frac{(PSO_3)^2}{(PSO_3)^2(PO_2)} = 3.76$$

$$(PSO_3)^2 = (3.76)(0.2)^2(0.5) = 0.0752$$
$$PSO_3 = 0.274$$

42. The mass of N_2O_4 would increase and NO_2 would decrease. The volume coefficient of the left side (1) is less than that of the right side (2), so a conversion of NO_2 to N_2O_4 would minimize the increase in pressure.

43. The mass of NO_2 would increase, and N_2O_4 would decrease. Because the forward reaction is endothermic,

$$N_2O_4 + \text{heat} \rightleftarrows 2\ NO_2$$

the conversion of N_2O_4 to NO_2 would absorb heat and minimize the increase in temperature.

44. The enthalpy of reaction is +33.7 kcal, so the reaction is endothermic.

$$\Delta H = \left[\overset{\text{MgO}}{(-602)} + 2\overset{\text{HCl}}{(-92)}\right] - \left[\overset{\text{MgCl}_2}{(-642)} + \overset{\text{H}_2\text{O}}{(-286)}\right] = -142\ \text{kJ}$$

45. The exothermic reaction releases 27.8 kilocalories of heat.

$$CaO(s) + H_2O(l) \rightarrow Ca(OH)_2(s)$$

$$H = \underbrace{[-987]}_{\text{product}} - \underbrace{[(-635) + (-286)]}_{\text{reactants}} = -66\ \text{kJ/mol}$$

$$H = (1.783\ \text{moles CaO})(-66\ \text{kJ/mol}) = 117.7\ \text{kJ}$$

46. The free energy change is 48.3 kcal; because this is positive, the reaction would not proceed.

$$\Delta G = \underbrace{\left[\overset{\text{NO}_2}{(52)} + \overset{\text{H}_2}{(0)}\right]}_{\text{products}} - \underbrace{\left[\overset{\text{NO}}{(87)} + \overset{\text{H}_2\text{O}(l)}{(-237)}\right]}_{\text{reactants}} = 202\ \text{kJ}$$

47. The temperature change reverses the reaction direction. From the standard values that are given, you can calculate that

$$\Delta H = 58.02\ \text{kJ}$$

$$\Delta S = 176.7\ \text{J/deg} = 0.1767\ \text{kJ/deg}$$

and then substitute these into

$$\Delta G = \Delta H - T\Delta S$$

At 25°C = 298 K, the free energy favors N_2O_4:

$$\Delta G = (58.02\ \text{kJ}) - (298)(0.1767\ \text{kJ/deg}) = 5.362\ \text{kJ}$$

At 100°C = 373 K, the free energy favors NO_2:

$$\Delta G = (58.02\ \text{kJ}) - (373)(0.1767\ \text{kJ/deg}) = -1.886\ \text{kJ}$$

Index

A

acetic acid, 102
acetylene, 58
acid ionization constant, 101
acids
  pH Scale, 98–100
  polyprotic, 104–106
  strong and weak, 100–103
actinides, 37
air, 84
alkali metals, 5, 28
alkaline solution, 99
alkanes, 56
alkenes, 57–58
alkyl halides, 61
alkynes, 58
alpha particles (α), 26
aluminum, 34
aluminum oxide, 15–16
aluminum-zinc cell, 119
ammonia, 63, 129
anions, 116
anode, 116
aqueous ammonia, 103
aqueous solution, 86
argon, 4–5, 34
aromatic compounds, 60
Arrhenius, Svante, 97
atomic masses, 5–7
atomic number, 8
atomic structure
  ions, 27–28
  isotopes, 23–25
  overview, 21
  radioactivity, 25–27
  subatomic particles, 22–23
atoms
  chemical compounds, 14–15
  chemical reactions, 18–19
  moles, 17–18
  overview, 13
  stoichiometry, 15–17
Avogadro, Amedeo, 75, 78
Avogadro's law, 78
Avogadro's number, 75
azimuthal quantum number, 32

B

balancing equations, 110–112
bases
  pH Scale, 98–100
  types of, 103–104
benzene, 60
beryllium, 22, 33
beta particles (β), 26
binary compounds, 89
boiling point, 55–56, 91–94
book resources, 161
boron, 23, 33
Boyle, Robert, 75
Boyle's law, 75–76
bromides, solubility, 87
bromine, 4–5
Brønsted, Johannes, 104

C

calcium, 34
calcium hydroxide, 103
calorie, 71, 148
carbon, 23, 33
carbon dioxide, phase diagram, 70
carbonated soda water, 84
carbonates, 88
carbonic acid, 102, 105
carboxylic acids, 103
cathode, 116
cations, 116
Celsius scale, 77
Chadwick, James, 23
Charles, Jacques, 77
Charles' law, 77–78
chemical bonding
  covalent bonds, 40–43
  hydrogen bond, 49–50
  ionic bonds, 43–46
  overview, 39
  polar bonds, 47–49
chemical energy, 141
chlorates, 87
chloric acid, 102
chlorides, 87

chlorine
  defined, 4–5
  electron configurations and oxidation numbers, 34
  isotopes, 24
  valence electrons, 41
cis structural configuration, 61–62
combining masses, 6
compounds
  chemical, 14–15
  enthalpies of, 138
  stoichiometric, 15–16
concentration units, 84–86
concentrations, equilibrium, 128–132
condensation, 73
condensed structural formulas, 56
congeners, periodic table, 9
conjugate acid-base pair, 104
copper, 27
Coulomb's law, 44–45
covalent bonds, 40–43
critical point, phase diagram, 70
crystalline structure, 66
crystallization, 73
cyclohexane, 59

D

Dalton, John, 6, 23
dilute solution, 87
dinitrogen pentoxide, 14–15
dinitrogen trioxide, 15
dissociation, 89–90
double covalent bond, 43
d-type subshells, 36

E

electrochemistry
  electrochemical cells, 115–117
  electrode potential, 117–123
electrodes, 116
electrolysis, 115, 120–121
electrolytes, 116
electrolytic cells, 120
electromotive force (EMF), 117–123
electron configurations
  inert gases, 40
  orbitals, 31–33
  overview, 31
  periodic table, 35–37
  valence, 33–35
electron transfer, 109–110
electronegativity, 47–49
elements
  atomic masses, 5–7
  discovery, 3–4
  periodic table, 7–9
  similarity, 4–5
EMF (electromotive force), 117–123
endothermic reactions, 134
energy, 141
enthalpy, 138–141
entropy, 141–144
equations, balancing, 110–112
equilibrium
  concentrations, 128–132
  defined, 98
  Le Chatelier's Principle, 132–135
  overview, 125
  reaction directions, 126–128
equilibrium constants, 130
ethylene, 57, 61
exothermic reactions, 134

F

Faraday, Michael, 121
faraday unit, 122
fluorine
  defined, 4–5
  electron configurations and oxidation numbers, 34
  valence electrons, 40–41
free energy, 137, 144–149
freezing point, 77, 90–94
freezing point depression, 90
functional groups, organic compounds, 62–63

G

gases
  Avogadro's law, 78
  Boyle's law, 75–76
  Charles' law, 77–78
  defined, 66
  halogens, 4–5
  ideal, 79
  Ideal Gas Equation, 79–80
  inert, 4–5, 40
  overview, 75
geometric isomers, 61–62
Gibbs, Josiah, 144

Gibbs free energy, 137, 144–149
gram formula mass, 17
ground state, orbitals, 31
groups, periodic table, 9

H

halogens, 4–5, 28
heat, 71
heat capacities, 71–73
heat flow, 139
heat of reaction, 138
helium
  defined, 4–5
  electron configurations and oxidation numbers, 33
  nuclear structure of, 22
hexane, 59
hydrocarbons, 53, 56–60
hydrochloric acid, 102
hydrogen
  electron configurations and oxidation numbers, 33
  isotopes, 25
  nuclear structure of, 22
hydrogen bond, 49–50
hydrogen chloride molecule, 42
hydroxides, 88

I

Ideal Gas Equation, 79–80
ideal gases, 79
independent variables, phase diagrams, 66–67
inert gases, 4–5, 40
International Union of Pure and Applied Chemistry (IUPAC), 11, 58
Internet resources, 161–162
iodides, 87
iodine, 4–5, 8
ionic bonds, 43–46
ion-product constant, 98
ions
  defined, 21
  electron configurations of, 35
  general discussion, 27–28
isoelectronic, 35, 46
isomers, 53
isothermal transformations, 72
isotopes, 23–25
IUPAC (International Union of Pure and Applied Chemistry), 11, 58

J

joule, 71, 148

K

Kelvin scale, 77
kilojoule, 71
krypton, 4–5

L

lanthanides, 37
Lavoisier, Antoine, 4
law of conservation of mass, 19
Le Chatelier, Henri, 132
Le Chatelier's Principle, 132–135
Lewis, Gilbert, 40
Lewis symbols, 39
liquids, 66
lithium
  defined, 4–5
  electron configurations and oxidation numbers, 33
  nuclear structure of, 22
litmus paper, 97
Lowry, Thomas, 104

M

magnesium, 34
mass number, 22
mass percents, 84
matter, states of. *See* states of matter
Mendeleyev, Dmitri, 7
metals, 8, 43. *See also* specific metals by name
methane, 61
molality, 86
molar heat capacity, 71
molarity, 86
mole fraction, 85
molecular formula, 54–55
molecules, 14
moles, 17–18

N

negative entropy of reaction, 144
negative ions, 27
neon, 4–5, 34
neutralization process, 98

neutrons, 22, 23
nitrates, 87
nitric acid, 102
nitric oxide, 15
nitrogen, 23, 34
nitrogen dioxide, 15
nitrous oxide, 14–15
noble gases
  defined, 4–5
  electron configurations, 40
nonmetals, 8, 44
nonmolecular substances, 17
nucleons, 23
nucleus, atom, 22

O

orbitals, 31–33
organic compounds
  with additional elements, 60–63
  hydrocarbons, 56–60
  overview, 53
  structural formulas, 54–56
oxidation numbers, 33–34
oxidation-reduction reactions
  balancing equations, 110–112
  electron transfer, 109–110
  overview, 107
  oxidation numbers, 108–109
oxides
  electron transfer, 109
  nitrogen, 14–15
oxidizing agent, 110
oxygen, 23, 34

P

Pascal unit, 76
Pauling, Linus, 47
periodic table, 7–9, 35–37
periods, periodic table, 7
pH Scale, 98–100
phase diagrams, 66–70
phosphates, 88
phosphoric acid, 102, 105
phosphorus
  electron configurations and oxidation numbers, 34
  valence electrons, 41
planar molecules, 61
polar bonds, 47–49
polyprotic acids, 104–106
positive entropy change, 143
positive ions, 27
potassium, 4–5, 34
potassium hydroxide, 103
pressure, 66–70, 75–76
Priestley, Joseph, 3–4
propane, structural formula, 54
propene, 57–58
protons, 22
p-type subshells, 36
pure covalent bonding, 48

Q

Quantum theory, orbitals, 31

R

radioactivity (radioactive decay), 25–27
rare earths, 37
reaction directions, equilibrium, 126–128
reactions. *See* specific reaction by name
redox reactions. *See* oxidation-reduction reactions
reducing agent, 110
resources, 161–162
Rutherford, Ernest, 22

S

salt formers, 4–5, 28
saturated hydrocarbons, 59
saturated solutions, 87
seawater, 84
shells, 31
silicon, 34
sodium, 4–5, 34
sodium chloride, 100
sodium hydroxide, 103
solids, 66
solubility, 87–91
solutes, 84
solutions
  boiling point, 91–94
  concentration units, 84–86
  freezing point, 91–94
  overview, 83–84
  solubility, 87–91
solvents, 84
specific heat capacity, 71
standard enthalpies, 139–140
standard entropies, 143

standard free energies of formation, 145
standard temperature and pressure (STP), 17
states of matter
gases, 66
heat capacities, 71–73
liquids, 66
overview, 65
phase diagrams, 66–70
solids, 66
transformations, 72–73
stoichiometric compounds, 15–16
stoichiometry, 15–17
STP (standard temperature and pressure), 17
strong acids, 100–103
strong electrolytes, 100
structural formulas, organic compounds, 54–56
s-type subshells, 35–36
subatomic particles, 22–23
sublimation, 67
subshells
defined, 32–33
d-type, 36–37
p-type, 36
s-type, 35–36
sulfates, 87
sulfides, 88
sulfites, 88
sulfur, 34
sulfuric acid, 102, 105
sulfurous acid, 101–102, 105
supersaturated solutions, 87

T

tellurium, 8
temperature, 66–70, 77
thermodynamics
energy, 141
enthalpy, 138–141
entropy, 141–144
Gibbs free energy, 144–149
overview, 137–138
Thomson, Joseph, 27
trans structural configuration, 61–62
transformations, states of matter, 72–73
transition metals, 36
transmutation, 26
triple covalent bond, 43
triple point, phase diagram, 70

U

unsaturated hydrocarbons, 59
unsaturated solutions, 87
uranium, 26

V

valence electrons, 33–35, 40–41
voltaic cells, 116–117, 120
voltmeter, 116
volume, 76–77

W

weak acids, 100–103
weak electrolytes, 101

X

xenon, 4–5

Z

zinc-copper cell, 119

Notes

Notes

Notes

Made in United States
North Haven, CT
10 November 2026

22084520R10104